American Book Company

Meeting Standards,
Exceeding Expectations

D1514736

Dear Educator,

Thank you for your interest in American Book Company's state-specific test preparation resources. We commend you for your interest in pursuing your students' success. Feel free to contact us with any questions about our books, software, or the ordering process.

Our Products Feature	Your Students Will Improve
Multiple-choice and open-ended diagnostic tests	Confidence and mastery of subjects
Step-by-step instruction	Concept development
Frequent practice exercises	Critical thinking
Chapter reviews	Test-taking skills
Multiple-choice practice tests	Problem-solving skills

American Book Company's writers and curriculum specialists have over 100 years of combined teaching experience, working with students from kindergarten through middle, high school, and adult education.

Our company specializes in effective test preparation books and software for high stakes graduation and grade promotion exams across the country.

How to Use This Book

Each book:

*contains a chart of standards which correlates all test questions and chapters to the state exam's standards and benchmarks as published by the state department of education. This chart is found in the front of all preview copies and in the front of all answer keys.

*begins with a full-length pretest (diagnostic test). This test not only adheres to your specific state standards, but also mirrors your state exam in weights and measures to help you assess each individual student's strengths and weaknesses.

*offers an evaluation chart. Depending on which questions the students miss, this chart points to which chapters individual students or the entire class need to review to be prepared for the exam.

*provides comprehensive review of all tested standards within the chapters. Each chapter includes engaging instruction, practice exercises, and chapter reviews to assess students' progress.

*finishes with two full-length practice tests for students to get comfortable with the exam and to assess their progress and mastery of the tested standards and benchmarks.

While we cannot <u>guarantee</u> success, our products are designed to provide students with the concept and skill development they need for the graduation test or grade promotion exam in their own state. We look forward to hearing from you soon.

Sincerely,

The American Book Company Team

PO Box 2638 ★ Woodstock, GA 30188-1383 ★ Phone: 1-888-264-5877 ★ Fax: 1-866-827-3240

Chart of Standards

Standard	Chapter Number	Diagnostic Test Part 1 Question #	Diagnostic Test Part 2 Question #	Practice Test 1 Part 1 Question #	Practice Test 1 Part 2 Question #	Practice Test 2 Part 1 Question #	Practice Test 2 Part 2 Question #
Number and Operations							
M2N1a	1	1	32	1	31	2	32
M2N1b	1	5	31	2	32	1	31
M2N1c	6	19	49	19	48	19	49
M2N2a	2	2	33	3	33	3	33
M2N2b	2	3	34	4	34	4	34
M2N2c	2						
M2N2d	2	4	35	6	57	6	35
M2N2e	2	6	36	5	35	5	36
M2N3a	3	7	37	7	36	7	37
M2N3b	3	9, 11	38, 39	8, 9, 11	37, 38	9, 11	38, 39, 41
M2N3c	3						
M2N3d	3	10, 12	40, 41	10, 12	39, 40	10, 12	40
M2N4a	4	13	43	13	42	15	43
M2N4b	4	14	42	14	41	13	42
M2N5	5				43, 44	14	44, 45
M2N5a	5	15	44, 46		45		46
M2N5b	5	16	45	16		16	
Measurement							
M2M1a	6	17	47	17	46	17	47
M2M1b	6	18	48	18	47	18	48
M2M1c	6	8		15		8	
M2M2	6	20	51	20	50	20	51
M2M3	6	21	50	21	49	21	50
Geometry							
M2G1	7	22, 23, 24	52, 53, 54	22, 23, 24	51, 52, 53	22, 23, 24	52, 53
M2G2	8	25, 26		25, 26		25, 26	
M2G2a	8		55, 56		54, 55		54, 55, 56
M2G2b	8		57		56		57
M2G3	8	27		27		27	
Data Analysis and Probability							
M2D1a	9						
M2D1b	9	28, 29, 30	58, 59, 60	28, 29, 30	58, 59, 60	28, 29, 30	58, 59, 60

MASTERING THE GEORGIA
2nd GRADE CRCT

IN

MATHEMATICS

Developed to the new Georgia Performance Standards!

ERICA DAY

COLLEEN PINTOZZI

TRISHA PASTER

American Book Company

P. O. Box 2638

Woodstock, Georgia 30188-1383

Toll Free 1 (888) 264-5877 Phone (770) 928-2834

Toll Free Fax 1 (866) 827-3240

WEB SITE: www.americanbookcompany.com

Acknowledgements

In preparing this book, we would like to acknowledge Mary Stoddard and Eric Field for their contributions developing graphics and Camille Woodhouse for her contributions in editing and formatting for this book. We would also like to thank our many students whose needs and questions inspired us to write this text.

Contents

Contents

Preface

Mastering the Georgia 2nd Grade CRCT in Mathematics will help you review and learn important concepts and skills related to elementary school mathematics. First, take the Diagnostic Test beginning on page 1 of the book. To help identify which areas are of greater challenge for you, complete the evaluation chart with your instructor in order to help you identify the chapters which require your careful attention. When you have finished your review of all of the material your teacher assigns, take the progress tests to evaluate your understanding of the material presented in this book. **The materials in this book are based on the Georgia Performance Standards including the content descriptions for mathematics, which are published by the Georgia Department of Education. The complete list of standards is located in the Answer Key. Each question in the Diagnostic and Practice Tests is referenced to the standard, as is the beginning of each chapter.**

This book contains several sections. These sections are as follows: 1) A Diagnostic Test; 2) Chapters that teach the concepts and skills for *Mastering the Georgia 2nd Grade CRCT in Mathematics*; and 3) Two Practice Tests. Answers to the tests and exercises are in a separate manual.

ABOUT THE AUTHORS

Erica Day has a Bachelor of Science Degree in Mathematics and is working on a Master of Science Degree in Mathematics. She graduated with high honors from Kennesaw State University in Kennesaw, Georgia. She has also tutored all levels of mathematics, ranging from high school algebra and geometry to university-level statistics, calculus, and linear algebra.

Colleen Pintozzi has taught mathematics at the middle school, junior high, senior high, and adult level for 22 years. She holds a B.S. degree from Wright State University in Dayton, Ohio and has done graduate work at Wright State University, Duke University, and the University of North Carolina at Chapel Hill.

Trisha Paster earned her Masters of Science degree in Education with a focus in Math and Science Education from the University of Central Florida. She has been writing and editing textbooks ranging in grades kindergarten–6 for nearly a decade. Before stepping into the publishing world, she taught mathematics at the middle school level and all core subjects at the elementary school level. She has a real passion for curriculum development and an extensive background in educational assessment.

Diagnostic Test

Part 1

1. Which number sentence represents 242?

    A $2 + 4 + 2$

    B $200 + 400 + 2$

    C $200 + 40 + 2$

M2N1a

2. What is the difference?

$$\begin{array}{r} 80 \\ -56 \\ \hline \end{array}$$

    A 24

    B 34

    C 136

M2N2a

3. Seth worked out this problem:

    $49 + 32 = 81$

Which of these can he use to check his work?

    A $81 - 32$

    B $81 + 49$

    C $49 - 32$

M2N2b

4. What is the addition property?

    $65 + 41 = 41 + 65$

    A Associative

    B Commutative

    C Identity

M2N2d

5. Madison has 902 stickers. Which model represents the number 902?

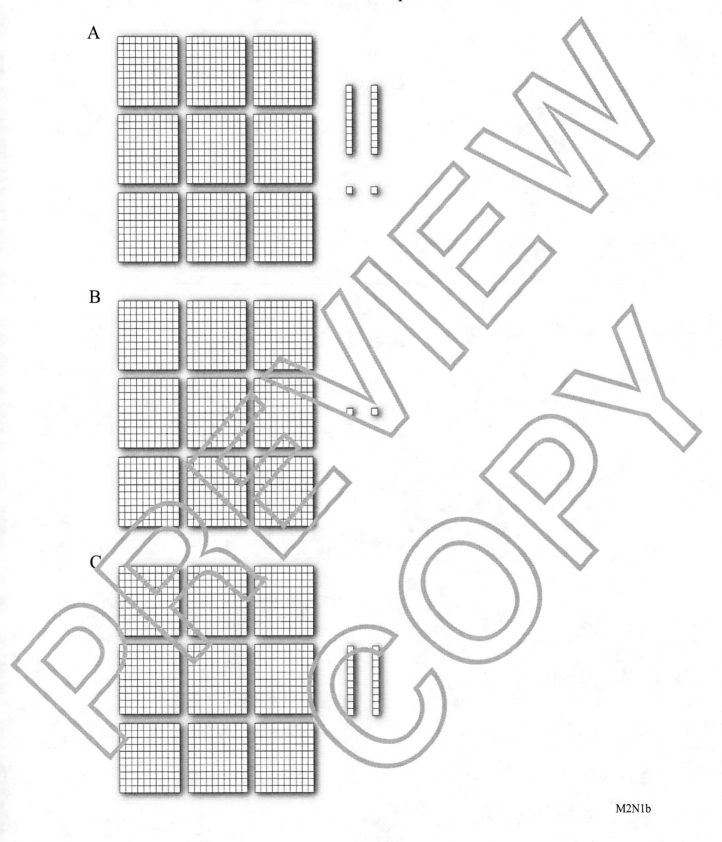

M2N1b

6. Which is the best estimate? Round to the nearest ten.

$$\begin{array}{r} 57 \\ -15 \\ \hline \end{array}$$

A 40

B 80

C 72

7. Which multiplication sentence relates to this addition sentence?

$2 + 2 + 2 = 6$

A $2 \times 2 =$

B $3 \times 2 =$

C $6 \times 2 =$

8. Which tool would you use to find the length of a kitchen table?

A ruler

B yardstick

C thermometer

9. What is the product?

$8 \times 3 =$

A 8

B 10

C 24

10. How many equal groups of 4 can be made from the robots below?

A 2

B 3

C 4

11. Which multiplication sentence matches the number line?

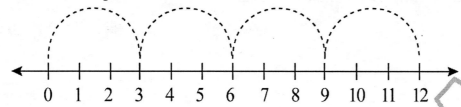

A $4 \times 4 =$

B $3 + 3 + 3 + 3 =$

C $3 \times 4 =$

12. What is the missing factor?

$24 - 6 = 18$ $18 - 6 = 12$ $12 - 6 = 6$ $6 - 6 = 0$ _____ $\times 6 = 24$

A 2

B 3

C 4

13. Which model shows an equal amount to the model below?

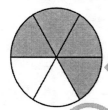

A

B

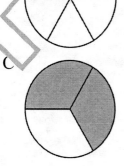

C

14. There are 4 kittens in the basket. All 4 of them are sleeping. Which fraction shows how many kittens are sleeping?

A $\dfrac{4}{2}$

B $\dfrac{1}{4}$

C $\dfrac{4}{4}$

15. Which symbol should be used to compare the numbers?

424 ☐ 434

A >
B <
C =

16. Cody decided to walk 140 minutes each week. This week he has walked 88 minutes so far. How many minutes of walking does he have left?

A 12
B 40
C 52

17. What is the length to the nearest centimeter?

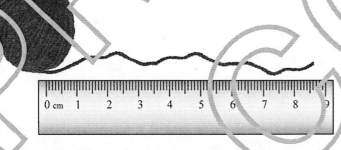

A 8 centimeters
B 9 centimeters
C 10 centimeters

18. Which length listed is the best estimate of the height of the door?

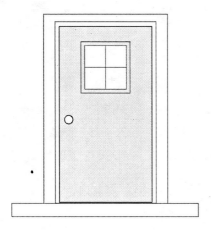

A 7 inches

B 7 feet

C 7 yards

19. Hannah gives the clerk a one dollar bill to buy a drink for $0.65. How much change should she receive?

A $0.35

B $0.65

C $1.65

20. What number should the minute hand be at when it is 3:10?

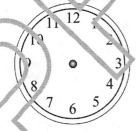

A 2

B 3

C 12

21. What is the temperature?

    A 45 degrees

    B 50 degrees

    C 55 degrees

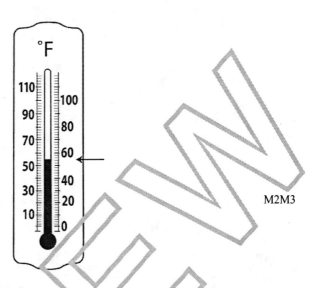

M2M3

22. How many vertices does this plane shape have?

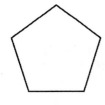

    A 3

    B 4

    C 5

M2G1

23. How many edges does a trapezoid have?

    A 3

    B 4

    C 5

M2G1

24. What angle measures less than 90°?

    A acute

    B right

    C obtuse

M2G1

25. How many edges does this solid have?

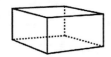

    A 0
    B 6
    C 12

M2G2

26. Which solid can roll?

    A cube
    B cylinder
    C pyramid

M2G2

27. Which new solid can be made with these two solids?

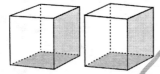

    A cube
    B pyramid
    C rectangular prism

M2G3

28. Which is the least favorite place to swim?

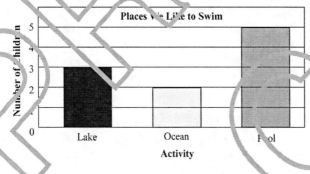

    A lake
    B ocean
    C pool

M2D1b

29. How many children in total like banana and strawberry as favorite smoothies?

Smoothies We Like				
Banana	🍌	🍌	🍌	
Grape	🍇	🍇		
Strawberry	🍓	🍓	🍓	🍓

Key: Each picture = 2 children

A 6

B 8

C 14

M2D1b

30. The Venn diagram below shows different types of shapes. How many black shapes are there in all?

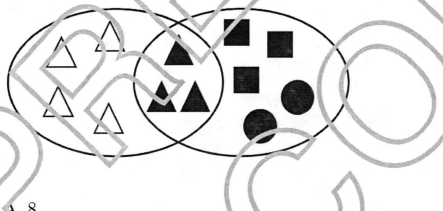

A 8

B 5

C 3

M2D1b

9

Part 2

31. Mrs. Annie has 568 books. Which model represents the number 568?

A

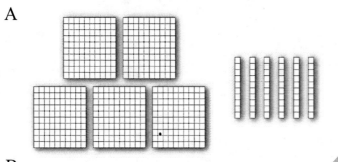

B

C

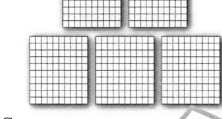

M2N1b

32. Which number sentence represents 819?

    A $800 + 100 + 9$
    B $800 + 19 + 9$
    C $800 + 10 + 9$

M2N1a

33. What is the sum?

$$\begin{array}{r} 809 \\ +148 \\ \hline \end{array}$$

  A 741
  B 947
  C 957

M2N2a

34. Ginny worked out this problem:

$$72 + 44 = 116$$

Which of these can she use to check her work?

A $72 - 44$
B $116 + 44$
C $116 - 44$

35. What is the addition property?

$$(7 + 2) + 4 = 7 + (2 + 4)$$

A Associative
B Commutative
C Identity

36. Which is the best estimate?

$$\begin{array}{r} 66 \\ -23 \\ \hline \end{array}$$

A 43
B 50
C 89

37. Which multiplication sentence relates to this addition sentence?

$$5 + 5 + 5 + 5 + 5 + 5 = 30$$

A $5 \times 5 =$
B $6 \times 5 =$
C $5 \times 30 =$

38. What is the product?

$$9 \times 5 =$$

A 27
B 36
C 45

39. Which multiplication sentence matches this array?

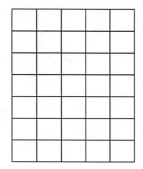

A $5 \times 5 =$

B $6 \times 5 =$

C $7 \times 5 =$

M2N3b

40. How many equal groups of 5 can be made with the cats shown below?

A 2

B 3

C 4

M2N3d

41. What is the missing factor?

$32 - 4 = 28$ $28 - 4 = 24$ $24 - 4 = 20$ $20 - 4 = 16$

$16 - 4 = 12$ $12 - 4 = 8$ $8 - 4 = 4$ $4 - 4 = 0$

$\underline{\hspace{1cm}} \times 4 = 32$

A 6

B 7

C 8

M2N3d

42. Which fraction represents the shaded part of the model?

A $\dfrac{1}{6}$

B $\dfrac{4}{6}$

C $\dfrac{6}{6}$

M2N4b

43. Leo ate $\dfrac{4}{6}$ of the granola bar. Which shows $\dfrac{4}{6}$ of the granola bar that is NOT shaded?

A

B

C

M2N4a

44. Which symbol should be used to compare the number of objects?

A >
B <
C =

M2N5a

45. Quincy has 6 more crackers than Greyson. Greyson has 8 crackers. Which comparison matches the information?

A $14 > 6 + 8$
B $14 < 6 + 8$
C $14 = 6 + 8$

46. What is the missing value?

$3 \times \boxed{} = 27$

A 7
B 8
C 9

47. What is the length to the nearest inch?

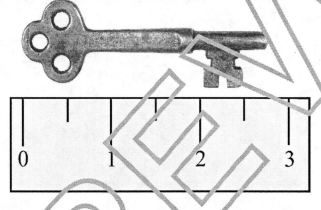

A 1 inch
B 2 inches
C 3 inches

48. Which length listed is the best estimate of the length of the shoe?

A 12 inches
B 12 feet
C 12 yards

49. Mrs. Levine bought a salad for $2.59 and a drink for $1.50. She gave the clerk $5.00. How much change should she receive back?

A $0.50

B $0.91

C $1.00

M2M1c

50. What is the temperature?

A 73 degrees

B 80 degrees

C 83 degrees

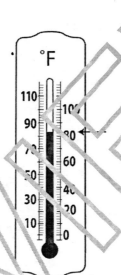

M2M3

51. What time does the clock show?

A 9:35

B 9:07

C 8:35

M2M2

52. Which plane shape listed has exactly 3 angles?

A rectangle

B square

C triangle

M2G1

53. What type of angle is shown?

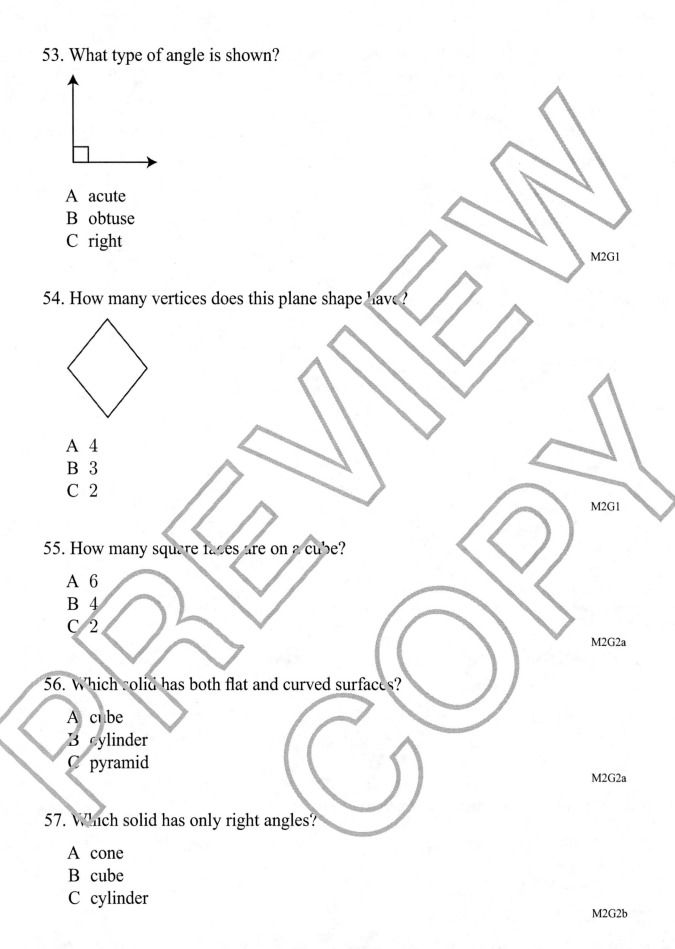

A acute
B obtuse
C right

M2G1

54. How many vertices does this plane shape have?

A 4
B 3
C 2

M2G1

55. How many square faces are on a cube?

A 6
B 4
C 2

M2G2a

56. Which solid has both flat and curved surfaces?

A cube
B cylinder
C pyramid

M2G2a

57. Which solid has only right angles?

A cone
B cube
C cylinder

M2G2b

16   Copyright ©American Book Company

58. How many children picked strawberry as a favorite smoothie?

Smoothies We Like					
Banana	🍌	🍌	🍌		
Grape	🍇	🍇			
Strawberry	🍓	🍓	🍓	🍓	

Key: Each picture = 2 children

A 4
B 5
C 8

59. How many children in all voted for a favorite place to swim?

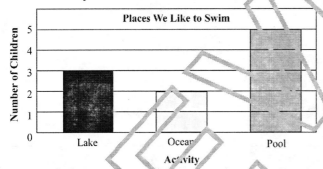

A 3
B 5
C 10

60. The Venn diagram shows different types of shapes. How many triangles are there in all?

A 7
B 4
C 3

Evaluation Chart for the Diagnostic Mathematics Test

Directions: On the following chart, circle the question numbers that you answered incorrectly. Then turn to the appropriate topics, read the explanations, and complete the exercises. Review the other chapters as needed. Finally, complete the *Mastering the Georgia 2nd Grade CRCT in Mathematics* Practice Tests to further review.

		Questions - Part 1	**Questions - Part 2**	**Pages**
Chapter 1:	Numbers	1, 5	31, 32	19–34
Chapter 2:	Addition and Subtraction	2, 3, 4, 6	33, 34, 35, 36	35–48
Chapter 3:	Multiplication	7, 9, 10, 11, 12	37, 38, 39, 40, 41	49–64
Chapter 4:	Fractions	13, 14	42, 43	65–81
Chapter 5:	Math Reasoning	15, 16	44, 45, 46	82–92
Chapter 6:	Measurement	8, 17, 18, 19, 20, 21	47, 48, 49, 50, 51	93–113
Chapter 7:	Plane Geometry	22, 23, 24	52, 53, 54	114–131
Chapter 8:	Solid Geometry	25, 26, 27	55, 56, 57	132–147
Chapter 9:	Graphs	28, 29, 30	58, 59, 60	148–164

Chapter 1
Numbers

This chapter covers the following Georgia Performance Standards:

| M2N | Number and Operations | M2N1a, b |

Naming Numbers in Word Form

You need to understand place value to name numbers. **Place value** is the value of the numeral based on its place in the number. One way to name a number is by using word form.

EXAMPLE: Use words to name the number 345.

Step 1: Look at the numeral in the largest place value.
3

Step 2: Write the numeral in words and then write the place value in words. This is a three, and it is in the hundreds place value.
Three hundred

Step 3: Now look at the next numeral in the largest place value
4

Step 4: Write the numeral in words and then write the place value in words. This is a four, and it is in the tens place value. A four in the tens place is called forty.
Forty

Step 5: Look at the next numeral in the largest place value.
5

Step 6: Write the numeral in words and add it to the forty. Place a dash between the tens and the ones place.
Forty-five

ANSWER: Three hundred forty-five

Write the numbers in word form.

1.   178

2.   62

3.   55

4.   238

5.   17

6.   80

7.   372

8.   901

9.   222

10.   31

Write the numbers in numerals.

11. sixty-seven

12. two hundred twenty-two

13. seventy

14. ninety-four

15. six hundred ten

16. ninety-eight

17. eight hundred eighteen

18. four hundred six

19. three hundred sixty-four

20. ten

Naming Numbers with Number Sentences

Another way you can name numbers is by making number sentences with the place value of each numeral.

EXAMPLE: Use number sentences to name 472.

Step 1: Look at the numeral in the largest place value. Add zeros to show the place value.
400

Step 2: Now look at the next numeral in the largest place value. Add zeros to show the place value.
70

Step 3: Now look at the next numeral in the largest place value. Since 2 is in the ones place, it has no zeros after it.
2

Step 4: Show the number in a number sentence.
400 + 70 + 2

ANSWER: 400 + 70 + 2

Write the numbers in number sentences.

1. 421

2. 34

3. 780

4. 46

5. 117

6. 908

7. 22

8. forty-eight

9. six hundred twelve

10. ninety-one

11. four hundred eight

12. seven hundred seventeen

You can also write numbers in number sentences by using both numerals and words.

EXAMPLE: Use number sentences to name 629.

Step 1:   Look at the numeral in the largest place value. Use numerals and words.
6 hundreds

Step 2:   Now look at the next numeral in the largest place value. Name the number.
twenty

Step 3:   Now look at the next numeral in the largest place value. Name the number.
9

Step 4:   Show the number in a number sentence.
6 hundreds + twenty + 9

ANSWER: 6 hundreds + twenty + 9

Write the numbers in number sentences using numerals and words.

1.   117

2.   23

3.   644

4.   27

5.   107

6.   888

7.   62

8.   491

9.   78

10.   164

Naming Numbers Using Models and Drawings

Models and drawings are another way to represent numbers.

EXAMPLE: Use the model to name the number.

Step 1: Look at the model. What units are shown?

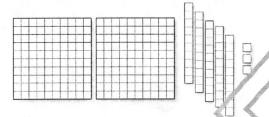

There are hundreds, tens, and ones shown.

Step 2: Start with the largest place value. Name it.

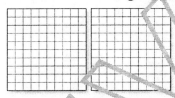

There are two hundreds shown. 200

Step 3: Look at the next largest place value shown, and name it.

There are five tens shown. 50

Step 4: Now look at the next largest place value shown, and name it.

There are three ones shown. 3

Step 5: Add all the numbers.
$200 + 50 + 3 = 253$

ANSWER: 253

Use the models to name the numbers.

1.

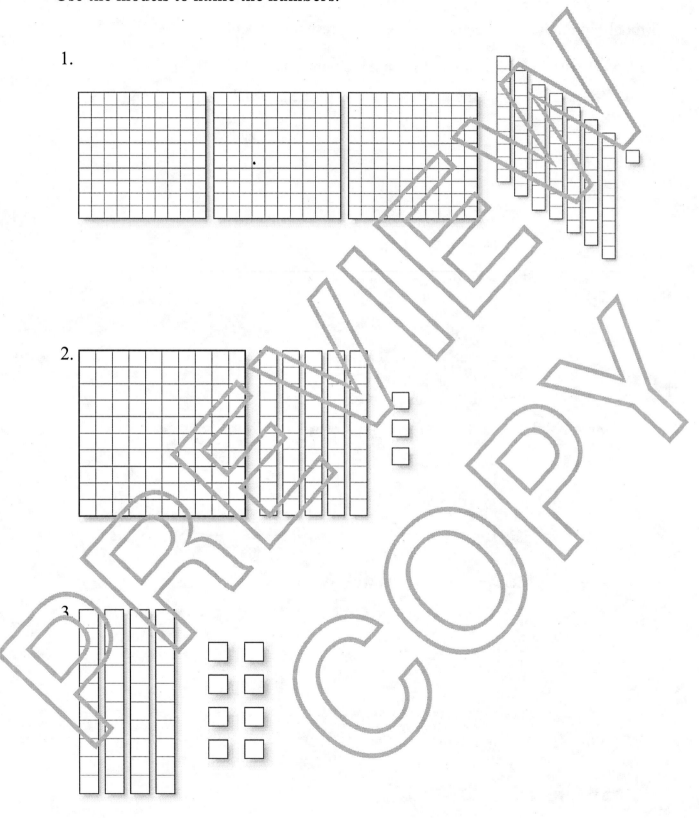

2.

3.

Copyright © American Book Company

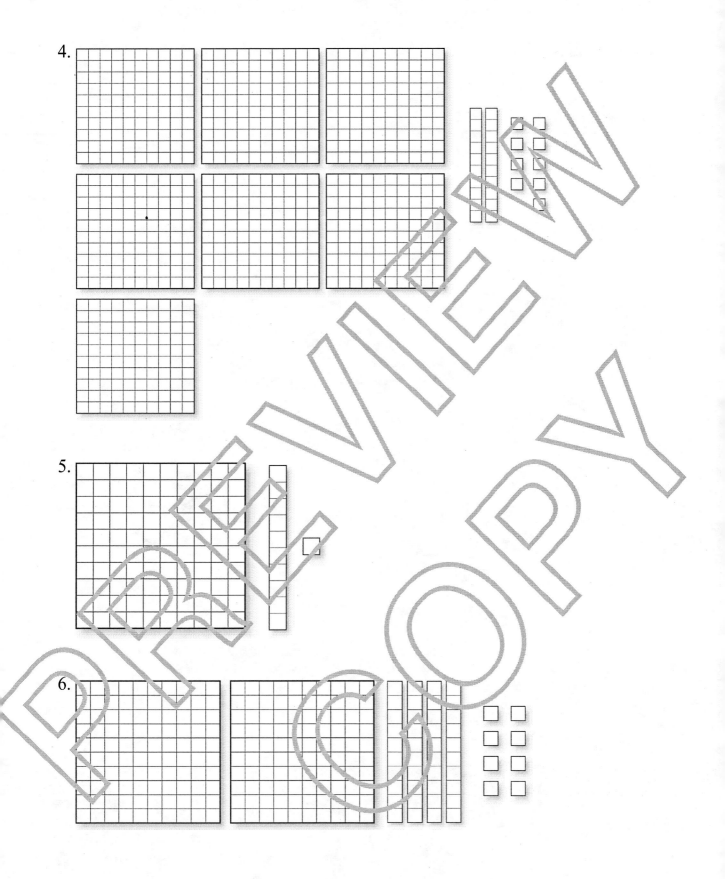

4.

5.

6.

Draw models to show the numbers.

7.   313

8.   25

9.   106

Naming Numbers Using Place Value Charts

Each section on a place value chart is worth a different amount. Each section is 10 times the value as the column to the right of it.

Thousands	Hundreds	Tens	Ones
1,000	100	10	1

Ones Place: 1

Tens Place: 10 times as much as the ones place

Hundreds Place: 10 times as much as the tens place

Thousands Place: 10 times as much as the hundreds place

EXAMPLE: Draw a model on the place value chart to represent the number 23.

Step 1:   Look at the numeral in the largest place value. Draw a model of the number on the place value chart. Write the numeral in the correct section of the place value chart.

Thousands	Hundreds	Tens	Ones
		2	

Step 2:   Look at the numeral in the next largest place value. Draw a model of the number on the place value chart. Write the numeral in the correct section of the place value chart.

Thousands	Hundreds	Tens	Ones
		2	3

Draw models and write numerals on the place value chart to represent the numbers.

1.   313

Thousands	Hundreds	Tens	Ones

2.   63

Thousands	Hundreds	Tens	Ones

Chapter 1 Review

Write the numbers in word form.

1.   258

2.   91

3.   267

4.   490

Write the numbers in numerals.

5. seventy

6. eight hundred twenty-nine

7. three hundred eight

8. sixty-two

Write the numbers in number sentences.

9. four hundred sixteen

10.   61

11.   910

Write the numbers in number sentences using numerals and words.

12.   113

13.   68

14.   454

Use the models to name the numbers.

15.

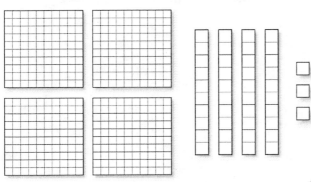

16.

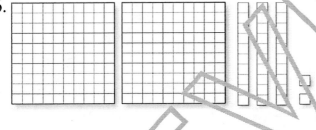

Draw models and write numerals on the place value chart to represent the number.

17. 223

Thousands	Hundreds	Tens	Ones

Chapter 1 Test

1. Madison has 607 stickers. Which model represents the number 607?

A

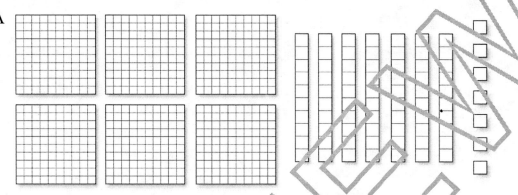

B

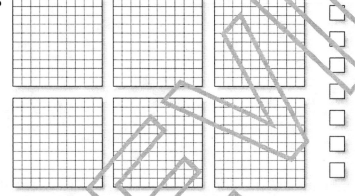

C

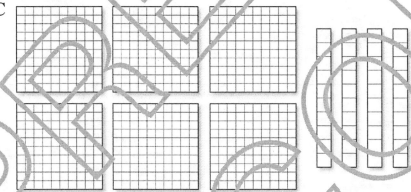

2. Which number sentence represents 121?

  A $1 + 2 + 1$

  B $100 + 200 + 1$

  C $100 + 20 + 1$

3. Which shows the number 390 in word form?

   A three hundred nine

   B three hundred ninety

   C thirty-nine tens nine

4. Which number sentence represents 718?

   A $700 + 18$ tens

   B 7 hundreds $+ 10 +$ eight

   C 7 hundreds $+$ eighty-one

5. The model represents which number?

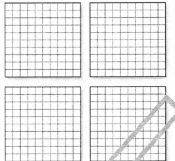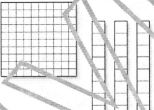

   A 563

   B 506

   C 5560

6. Which number represents seven hundred sixteen in numerals?

   A 7,016

   B 716

   C 760

7. Which number sentence represents 999?

   A $900+$ ninety $+ 9$

   B 9 hundreds $+$ 9 ones

   C $900+$ ninety

8. Which number sentence represents 203?

   A 2 hundred + 3

   B 230 + three

   C twenty + three

9. Which number sentence represents 516?

   A 51 + 6

   B 500 + 100 + 16

   C 500 + 10 + 6

10. Which number represents four hundred two in numerals?

   A 42

   B 420

   C 402

11. Which number sentence represents 71?

   A 70 + 1

   B 700 + 100

   C 700 + 10

12. The model represents which number?

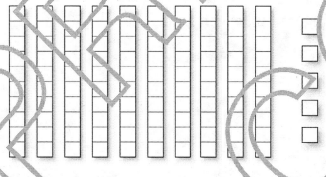

   A 1,005

   B 15

   C 105

Chapter 2
Addition and Subtraction

This chapter covers the following Georgia Performance Standards:

M2N	Number and Operations	M2N2a, b, c, d, e

Adding Whole Numbers

$$\boxed{\bigcirc{2} \; + \; \bigcirc{1} \; = \; \boxed{3}}$$

The numbers with circles around them are addends. **Addends** are numbers that are added together. The number with a square around it is the sum. The **sum** is the answer to an addition problem.

At times, you will need to regroup when you add. **Regrouping** in addition happens when the addends together are larger than the number 9.

Caroline has 22 stickers in her album. Her mom gives her 18 more stickers. How many stickers does Caroline have in all?

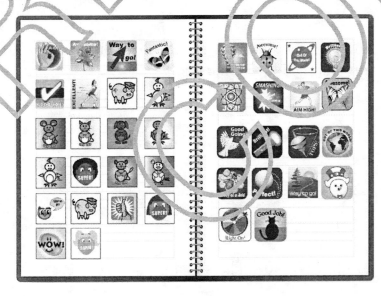

EXAMPLE: Find the sum. $22 + 18 =$

Step 1:   Write the addition problem up and down.

$$\begin{array}{r} 22 \\ + 18 \\ \hline \end{array}$$

Step 2:   Start in the ones column and add the numbers together.

$$\begin{array}{cc} \text{Tens} & \text{Ones} \\ 22 \\ + 18 \\ \hline \end{array}$$

$2 + 8 = 10$ 10 is more than 9, so you must regroup.

Step 3:   Regroup.

$$\begin{array}{cc} \text{Tens} & \text{Ones} \\ 1 \\ 22 \\ + 18 \\ \hline 0 \end{array}$$

Step 4:   Now add the numbers in the tens column. Don't forget the 1 that you regrouped from the one's place.

$$\begin{array}{cc} \text{Tens} & \text{Ones} \\ 1 \\ 22 \\ + 18 \\ \hline 40 \end{array}$$

ANSWER: Caroline has 40 stickers in all.

Find the sum. Regroup if needed.

1. $\begin{array}{r} 672 \\ +19 \\ \hline \end{array}$

2. $55 + 96 =$

3. $\begin{array}{r} 90 \\ +43 \\ \hline \end{array}$

4. $\begin{array}{r} 19 \\ +65 \\ \hline \end{array}$

5. $\begin{array}{r} 44 \\ +12 \\ \hline \end{array}$

6. $712 + 838 =$

7. $99 + 11 =$

8. $\begin{array}{r} 17 \\ +71 \\ \hline \end{array}$

9. $\begin{array}{r} 522 \\ +124 \\ \hline \end{array}$

10. $\begin{array}{r} 84 \\ +37 \\ \hline \end{array}$

11. $101 + 209 =$

12. $34 + 92 =$

13. $155 + 76 =$

14. $\begin{array}{r} 412 \\ +363 \\ \hline \end{array}$

15. $\begin{array}{r} 45 \\ +832 \\ \hline \end{array}$

Subtracting Whole Numbers

The **difference** is the answer to a subtraction problem. You subtract the bottom number from the top number to find the difference. If the bottom number is larger than the top number, you will need to regroup before you can subtract.

Noah has 62 toy cars. He gave 13 to his cousin. How many toy cars does Noah have left?

EXAMPLE: Find the difference. $62 - 13 =$

Step 1:   Write the subtraction problem up and down. The original number listed goes on the top.

$$\begin{array}{r} 62 \\ -13 \\ \hline \end{array}$$

Step 2:   Start in the ones column and subtract the number on the bottom from the number on the top.

$$\begin{array}{r} \overset{\text{Tens Ones}}{62} \\ -13 \\ \hline \end{array}$$

Since 3 is larger than 2, you will need to regroup. (borrow)

Step 3:   Regroup. (Borrow 10 from the tens column.)

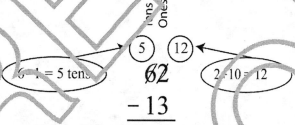

Step 4:   Now go back and subtract 3 from 12.

$$\begin{array}{r} \overset{(5)\;(12)}{\cancel{62}} \\ -13 \\ \hline \end{array}$$

Step 5:   Subtract the numbers in the tens place.

$$\begin{array}{r} 62 \\ -13 \\ \hline 49 \end{array}$$

ANSWER: Noah has 49 toy cars left.

Find the difference. Regroup if needed.

1. $\begin{array}{r} 515 \\ -12 \\ \hline \end{array}$

2. $92 - 18 =$

3. $\begin{array}{r} 60 \\ -52 \\ \hline \end{array}$

4. $\begin{array}{r} 188 \\ -144 \\ \hline \end{array}$

5. $\begin{array}{r} 56 \\ -19 \\ \hline \end{array}$

6. $712 - 638 =$

7. $87 - 9 =$

8. $\begin{array}{r} 99 \\ -61 \\ \hline \end{array}$

9. $\begin{array}{r} 432 \\ -124 \\ \hline \end{array}$

10. $\begin{array}{r} 50 \\ -27 \\ \hline \end{array}$

11. $100 - 83 =$

12. $56 - 12 =$

13. $222 - 33 =$

14. $\begin{array}{r} 876 \\ -543 \\ \hline \end{array}$

15. $\begin{array}{r} 190 \\ -67 \\ \hline \end{array}$

Using Addition and Subtraction to Check Problems

Addition is the opposite operation of subtraction. Subtraction is the opposite operation of addition.

If you find the sum of a problem, you can check it by using subtraction.

EXAMPLE: Use subtraction to check the sum.

Step 1:   Take the sum and write it as the top number.

$$
\begin{array}{r} 22 \\ +17 \\ \hline 39 \end{array} \qquad \nearrow \; 39
$$

Step 2:   Now subtract either of the addends from the sum.

$$
\begin{array}{r} 22 \\ +17 \\ \hline 39 \end{array} \qquad \begin{array}{r} 39 \\ -22 \\ \hline \end{array}
$$

Step 3:   The difference should match the other addend.

$$
\begin{array}{r} 22 \\ +17 \\ \hline 39 \end{array} \qquad \begin{array}{r} 39 \\ -22 \\ \hline 17 \end{array}
$$

At times, the difference might not match the other addend. If this happens, review both problems to find the mistake.

Use the opposite operation to check the answers. Write whether the operation is "correct" or "not correct."

1.
$$\begin{array}{r} 412 \\ +23 \\ \hline 435 \end{array}$$

6. $784 + 132 = 916$

11. $100 - 76 = 24$

2. $61 + 99 = 150$

7. $19 + 21 = 40$

12. $42 - 36 = 14$

3.
$$\begin{array}{r} 46 \\ +27 \\ \hline 63 \end{array}$$

8.
$$\begin{array}{r} 86 \\ -52 \\ \hline 34 \end{array}$$

13. $896 - 45 = 851$

4.
$$\begin{array}{r} 75 \\ +90 \\ \hline 165 \end{array}$$

9.
$$\begin{array}{r} 222 \\ -133 \\ \hline 111 \end{array}$$

14.
$$\begin{array}{r} 733 \\ -450 \\ \hline 323 \end{array}$$

5.
$$\begin{array}{r} 89 \\ +12 \\ \hline 101 \end{array}$$

10.
$$\begin{array}{r} 84 \\ -27 \\ \hline 63 \end{array}$$

15.
$$\begin{array}{r} 981 \\ -78 \\ \hline 917 \end{array}$$

Addition Properties

There are some basic rules, or properties, that are used for addition. These properties can make adding more simple.

Property	Model	Rule
Identity	$5 + 0 = 5$	When adding two numbers and one number is zero, the sum is equal to the other number.
Commutative	$4 + 2 = 2 + 4$	When adding two numbers, the sum will remain the same, no matter which order you use.
Associative	$(1 + 2) + 3 = 1 + (2 + 3)$	When adding three or more numbers, the sum is the same no matter which two numbers you add first.

Identify the property.

1. $5 + 0 = 5$

2. $(6 + 3) + 8 = 6 + (3 + 8)$

3. $80 + 4 = 4 + 80$

4. $34 + 78 = 78 + 34$

5. $8 + 0 = 8$

6. $4 + (9 + 28) = (4 + 9) + 28$

7. $62 + 7 = 7 + 62$

8. $(16 + 4) + 9 = 16 + (4 + 9)$

9. $0 + 25 = 25$

10. $8 + 9 = 9 + 8$

Addition and Subtraction Estimation

At times, you can estimate an answer instead of finding the exact answer. Estimating is another way to check your answers. There are different ways to estimate answers. You can change the number to a close number that is easy to use. This is called **rounding**. To round numbers to the nearest ten, round the numbers that end in 1 through 4 down to the next lower number that ends in 0. For example, 64 rounded to the nearest ten would be 60. Numbers that end in a digit of 5 through 9 should be rounded up to the next even ten. The number 78 rounded to the nearest ten would be 80.

EXAMPLE: Estimate $186 - 27 =$

Step 1: Round the numbers in the problem to the nearest ten.
186 will be rounded up to 190. 27 will be rounded up to 30.

$190 - 30 =$

Step 2: Subtract the new problem.

$$\begin{array}{r} 190 \\ -30 \\ \hline 160 \end{array}$$

ANSWER: $186 - 27$ is about 160.

To round numbers to the nearest hundred, round the numbers that end in 1 through 49 down to the next lower number that ends in 00. For example, 324 rounded to the nearest hundred would be 300. Numbers that have the last two digits of 50 through 99 should be rounded up to the next even hundred. The number 588 rounded to the nearest hundred would be 600.

EXAMPLE: Estimate $467 + 142 =$

Step 1: Round the numbers in the problem to the nearest hundred.
467 will be rounded up to 500. 142 will be rounded down to 100.

$500 + 100 =$

Step 2: Add the new problem.
$500 + 100 = 600$

ANSWER: $467 + 142$ is about 600.

Estimate the sum or difference. Round to the nearest ten.

1. $412 + 38 =$

2. $99 - 11 =$

3.
$$\begin{array}{r} 74 \\ -8 \\ \hline \end{array}$$

4.
$$\begin{array}{r} 459 \\ +165 \\ \hline \end{array}$$

5. $78 - 9 =$

Estimate the sum or difference. Round to the nearest hundred.

6.
$$\begin{array}{r} 672 \\ +119 \\ \hline \end{array}$$

7. $275 + 196 =$

8.
$$\begin{array}{r} 462 \\ -147 \\ \hline \end{array}$$

9.
$$\begin{array}{r} 397 \\ +431 \\ \hline \end{array}$$

10.
$$\begin{array}{r} 657 \\ -598 \\ \hline \end{array}$$

Chapter 2 Review

Find the sum. Regroup if needed.

1. 789
   $+102$

2. $67 + 115 =$

3. 60
   $+45$

Find the difference. Regroup if needed.

4. 260
   -148

5. $196 - 38 =$

6. 79
   -42

Use the opposite operation to check the answers. Write whether the operation is "correct" or "not correct".

7. 103
   $+89$
   $\overline{182}$

9. 43
   -27
   $\overline{24}$

8. $52 + 84 = 136$

10. $254 - 57 = 197$

Identify the property.

11. $67 + 23 = 23 + 67$

12. $(41 + 67) + 90 = 41 + (67 + 90)$

13. $48 + 0 = 48$

Estimate the sum or difference. Round to the nearest hundred.

14. 778
      $+484$
      ___

15. $211 - 146 =$

16. 397
      -284
      ___

Chapter 2 Test

1. Savannah worked out this problem:

$760 - 54 = 706$

Which of these can she use to check her work?

A $706 + 54$
B $706 - 54$
C $760 + 54$

2. What is the addition property?

$89 + 55 = 55 + 89$

A Associative
B Commutative
C Sum

3. Which is the best estimate? Round to the nearest hundred.

$$\begin{array}{r} 891 \\ +87 \\ \hline \end{array}$$

A 800
B 900
C $1,000$

4. What is the sum?

$$\begin{array}{r} 921 \\ +78 \\ \hline \end{array}$$

A 843
B 957
C 999

5. What is the addition property?

$7 + 0 = 7$

A Associative
B Commutative
C Identity

6. Jack worked out this problem:

$111 + 11 = 122$

Which of these can he use to check his work?

A $122 + 11$
B $111 - 11$
C $122 - 111$

7. What is the sum?

$$\begin{array}{r} 72 \\ +234 \\ \hline \end{array}$$

A 306
B 242
C 206

8. What is the difference?

$$\begin{array}{r} 90 \\ -88 \\ \hline \end{array}$$

A 2
B 12
C 18

9. What is the difference?

$$\begin{array}{r} 60 \\ -26 \\ \hline \end{array}$$

A 26
B 34
C 46

10. What is the best estimate? Round to the nearest ten.

$$\begin{array}{r} 67 \\ +25 \\ \hline \end{array}$$

A 40
B 80
C 100

Chapter 3
Multiplication

This chapter covers the following Georgia Performance Standards:

| M2N | Number and Operations | M2N3a, b, c, d |

Using Repeated Addition to Multiply

Multiplication is repeating addends over and over to find a product. The **product** is the answer to a multiplication problem.

Emma baked 2 muffins for each of her 3 friends. How many muffins did she bake in all?

EXAMPLE: Write a multiplication sentence to find the product.

Step 1: List the information you know from the problem.

Friend 1   Friend 2   Friend 3

Step 2: Write an addition sentence.

Friend 1   Friend 2   Friend 3

$$2 \quad + \quad 2 \quad + \quad 2 \quad = 6 \text{ muffins}$$

Step 3: Use the information to write a multiplication sentence.
There are 3 friends and each friend gets 2 muffins.

$$3(\text{friends}) \times 2(\text{muffins for each friend}) = 6(\text{muffins in all})$$

ANSWER: $3 \times 2 = 6$ Emma baked 6 muffins in all.

Use the addition sentence to write a multiplication sentence.

1. $3 + 3 + 3 = 9$

2. $4 + 4 + 4 = 12$

3. $2 + 2 + 2 + 2 = 8$

4. $5 + 5 = 10$

5. $3 + 3 + 3 + 3 + 3 = 15$

6. $7 + 7 = 14$

7. $1 + 1 + 1 + 1 + 1 + 1 = 6$

8. $6 + 6 + 6 = 18$

9. $8 + 8 = 16$

10. $2 + 2 + 2 + 2 + 2 = 10$

Using Arrays to Multiply

In the last section, you learned how multiplication is repeated addition. Arrays can help you make multiplication sentences from repeated addition. An **array** show amounts in rows and columns.

Below is an array with an apple in each section.

Each row has 4 apples. You could find the total amount of apples by adding each row.

$$4 + 4 + 4 =$$

Or you could make a multiplication sentence.

$$3 \text{ (rows)} \times 4 \text{ (apples in each row)} = 12$$

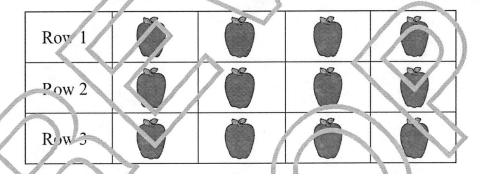

The numbers with circles around them are factors. **Factors** are numbers that are multiplied together. The number with a square around it is the product. The **product** is the answer to a multiplication problem.

$$\textcircled{3} \times \textcircled{4} = \boxed{12}$$

Use the array to write a multiplication sentence and find the product.

1.

2.

3.

4.

5.

6.

7.

8.

Using Skip Counting to Multiply

Another to way to find products is by skip counting.

EXAMPLE: Use skip counting to find the product.

Step 1: Look at the number line and count how many spaces are between each jump.

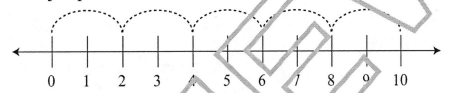

There are 2 spaces between each jump.

Step 2: Now count how many jumps in all.

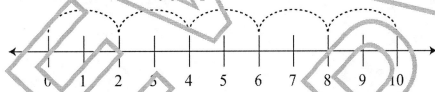

There are 5 jumps in all.

Step 3: The spaces between each jump and the number of jumps are the factors.
2 and 5 are factors.

Step 4: Use the factors to write a multiplication sentence and find the product.
$2 \times 5 = 10$

ANSWER: $2 \times 5 = 10$

Skip count to write a multiplication sentence and find the product.

1.
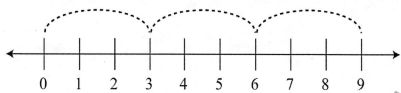

2.
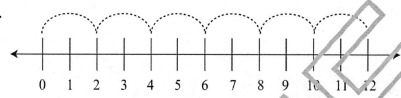

3.
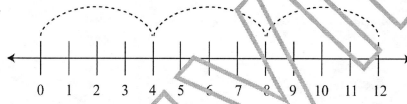

4.
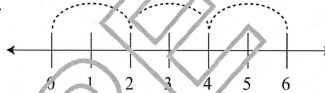

5.
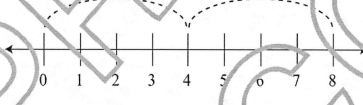

6.
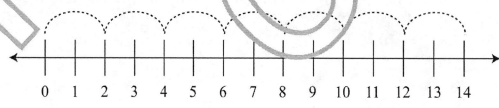

Constructing a Multiplication Table

Multiplication tables are also a way to find products. A multiplication table has repeated addition patterns.

EXAMPLE: Use the multiplication table to fill in the product of 3×2.

Step 1: Find the row with 3 in it.

×	0	1	2	3
0				
1				
2				
3				

Step 2: Now find the column with 2 in it.

×	0	1	2	3
0				
1				
2				
3				

Step 3: Write the product of 3×2 in the box where the row and column meet.

×	0	1	2	3
0				
1				
2				
3			6	

ANSWER: $3 \times 2 = 6$

Fill in the empty boxes of the multiplication table.

×	0	1	2	3	4	5	6	7	8	9
0	0		0	0		0	0			0
1		1			4				8	
2			4	6		10	12	14		18
3	0	3		9	12	15		21		27
4	0		8	12	16		24		32	
5		5			20	25	30		40	
6	0		12	18			36	42		54
7		7			28	35		49	56	
8			16	24			48		64	
9	0	9			36	45		63	72	81

Finding Products with the Multiplication Table

You filled in the multiplication table in the last section. You can now find products by using the multiplication table.

Use the multiplication table from 3.4 to find the product.

1. $6 \times 7 =$

2. $3 \times 8 =$

3. $9 \times 1 =$

4. $2 \times 8 =$

5. $4 \times 4 =$

6. $0 \times 3 =$

7. $5 \times 6 =$

8. $8 \times 5 =$

9. $2 \times 4 =$

10. $3 \times 7 =$

11. $4 \times 0 =$

12. $7 \times 8 =$

Using Repeated Subtraction to Divide

You can use repeated subtraction to divide. **Repeated subtraction** is when you keep subtracting the same number until you reach zero.

Joshua has 12 toy cars. He wants to give each of his friends 3 cars each. How many friends will Joshua give his cars to?

EXAMPLE: Use repeated subtraction to find the answer.

Step 1: List the information you know from the problem.

Joshua has 12 toy cars.

Step 2: Write subtraction sentences. Continue subtracting until you reach zero.

$$12 - 3 = 9 \qquad 9 - 3 = 6 \qquad 6 - 3 = 3 \qquad 3 - 3 = 0$$

Step 3: Count the amount of times you subtracted.

$$12 - 3 = 9 \qquad 9 - 3 = 6 \qquad 6 - 3 = 3 \qquad 3 - 3 = 0$$

There are 3 cars for each friend. There are 4 friends that will receive 3 cars each.

$$3(\text{cars}) \times 4(\text{friends}) = 12(\text{cars in all})$$

ANSWER: $3 \times 4 = 12$ Joshua will give 4 friends 3 cars each.

Use repeated subtraction to find the missing factor.

1. $16 - 4 = 12$ $12 - 4 = 8$ $8 - 4 = 4$ $4 - 4 = 0$

    ___ $\times 4 = 16$

2. $15 - 3 = 12$ $12 - 3 = 9$ $9 - 3 = 6$ $6 - 3 = 3$ $3 - 3 = 0$

    ___ $\times 3 = 15$

3. $21 - 7 = 14$ $14 - 7 = 7$ $7 - 7 = 0$

    ___ $\times 7 = 21$

4. $18 - 3 = 15$ $15 - 3 = 12$ $12 - 3 = 9$ $9 - 3 = 6$ $6 - 3 = 3$ $3 - 3 = 0$

    ___ $\times 3 = 18$

5. $18 - 2 = 16$ $16 - 2 = 14$ $14 - 2 = 12$ $12 - 2 = 10$ $10 - 2 = 8$

    $8 - 2 = 6$ $6 - 2 = 4$ $4 - 2 = 2$ $2 - 2 = 0$

    ___ $\times 2 = 18$

6. $9 - 9 = 0$

    ___ $\times 9 = 9$

Dividing Equally

Mrs. Jackson has 15 pencils. She wants to put them into 3 equal groups. How many pencils will be in each group?

EXAMPLE: Divide the items into equal groups.

Step 1: List the information you know from the problem.
Mrs. Jackson has 15 pencils.

Step 2: Make three circles to represent the groups.

Step 3: Start placing a pencil into each circle until all the pencils are used.

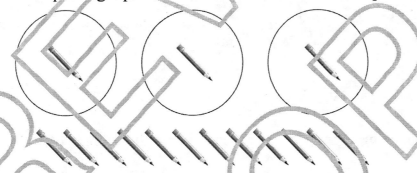

Step 4: Count how many pencils are equally divided into the 3 groups.

There are 5 pencils in each group.

$$5 \text{(pencils)} \times 3 \text{(groups)} = 15 \text{(pencils in all)}$$

ANSWER: $5 \times 3 = 15$. Mrs. Jackson can place 5 pencils each equally into the 3 groups.

Use the information to divide the items into equal groups.

1. How many equal groups of 2 can be made with the shells below?

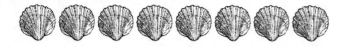

2. How many equal groups of 5 can be made with the seahorses below?

3. How many equal groups of 3 can be made with the sand dollars shown below?

4. How many equal groups of 4 can be made with the pails below?

5. How many equal groups of 9 can be made with the starfish shown below?

Chapter 3 Review

Use the addition sentence to write a multiplication sentence.

1. $4 + 4 + 4 + 4 = 16$

2. $5 + 5 + 5 = 15$

3. $2 + 2 + 2 + 2 = 8$

Use the array to write a multiplication sentence and find the product.

4.

5.

Skip count to write multiplication sentences and find the product.

6.

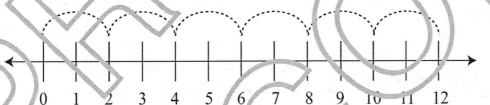

7.

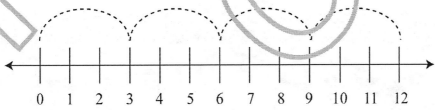

Fill in the Multiplication Table for problems 8–19.

×	0	1	2	3
0	0			
1	0			
2	0			
3	0			

Find the product.

20. $6 \times 7 =$ 21. $3 \times 8 =$ 22. $9 \times 1 =$

Use repeated subtraction to find the missing factor.

23. $12 - 4 = 8$ $8 - 4 = 4$ $4 - 4 = 0$

___ $\times 4 = 12$

24. $18 - 3 = 15$ $15 - 3 = 12$ $12 - 3 = 9$ $9 - 3 = 6$ $6 - 3 = 3$ $3 - 3 = 0$

___ $\times 3 = 18$

Use the information to divide the items into equal groups.

25. How many equal groups of 3 can be made with the snowmen below?

26. How many equal groups of 5 can be made with the scarves shown below?

Chapter 3 Test

1. Which multiplication sentence relates to this addition sentence?

$2 + 2 + 2 + 2 = 8$

A $2 \times 2 =$
B $8 \times 2 =$
C $4 \times 2 =$

2. Which multiplication sentence matches this array?

A $4 \times 4 =$
B $4 + 4 =$
C $16 \times 4 =$

3. Which multiplication sentence matches the number line?

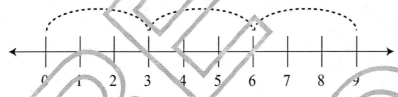

A $3 \times 3 =$
B $9 \times 3 =$
C $3 + 3 + 3 =$

4. How many equal groups of 6 can be made with the penguins below?

A 2
B 3
C 4

5. What is the product?

$8 \times 3 =$

A 5
B 11
C 24

6. What is the missing factor?

$21 - 7 = 14 \quad 14 - 7 = 7 \quad 7 - 7 = 0$

___ $\times 7 = 21$

A 2
B 3
C 4

7. How many equal groups of 4 can be made with the robots shown below?

A 2
B 4
C 5

8. What is the product?

$4 \times 5 =$

A 9
B 15
C 20

9. Which multiplication sentence relates to this addition sentence?

$9 + 9 = 18$

A $9 \times 2 =$
B $1 \times 9 =$
C $9 \times 9 =$

Chapter 4
Fractions

This chapter covers the following Georgia Performance Standards:

M2N	Number and Operations	M2N4a, b

Naming Fractions

Fractions are equal parts of a whole number. The number on top of a fraction is called the **numerator**. The number on the bottom of the fraction is called the **denominator**. Since denominator starts with "**D**," you can remember that it is the number on the fraction that goes **D**own.

$$\text{Numerator} \rightarrow \frac{2}{3} \leftarrow \text{Denominator}$$

Suppose your family is going to share one pizza. To share the pizza, you cut the pizza into pieces.

The total amount of equal pieces the pizza is cut into is the **denominator**. The amount you eat of the pizza would be the **numerator**. To be fair, all the pieces of the pizza must be cut into equal sizes.

Use N for numerator and D for denominator to label the parts of the fractions.

1. $\dfrac{6}{8}$

2. $\dfrac{1}{6}$

3. $\dfrac{9}{10}$

4. $\dfrac{1}{3}$

5. $\dfrac{4}{5}$

6. $\dfrac{1}{2}$

7. $\dfrac{8}{10}$

8. $\dfrac{4}{8}$

9. $\dfrac{2}{6}$

10. $\dfrac{4}{10}$

Modeling Fractions

EXAMPLE: Write the shaded part of the model as a fraction.

Step 1:   Look at the model. Count the amount of equal pieces in all.

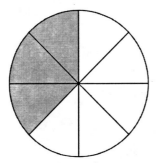

There are 8 equal pieces.

Step 2:   Place the total amount of equal pieces in the denominator.

$$\frac{\ }{8}$$

Step 3:   Now count the amount of shaded pieces.

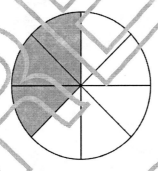

There are 3 shaded pieces.

Step 4:   Place the total amount of shaded pieces in the numerator.

$$\frac{3}{8}$$

ANSWER: $\frac{3}{8}$ of the model is shaded.

Write a fraction to name the shaded pieces of the model.

1.

2.

3.

4.

5.

6.

Fractions of a Whole

Hayden and 2 of his friends are going to share a fruit bar. They each want an equal piece of the fruit bar.

EXAMPLE: Divide a whole into equal pieces. Write a fraction to show the amount Hayden will get.

Step 1: Decide how many pieces the fruit bar needs to be divided into.

There is Hayden and his 2 friends. This means the fruit bar needs to be divided into 3 equal parts.

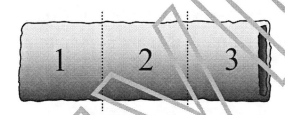

Step 2: Write a fraction.

How many pieces are there in all? 3

Place this number in the denominator.

$$\frac{}{3}$$

Step 3: How many pieces will Hayden get of the fruit bar? 1

Place this number in the numerator.

$$\frac{1}{3}$$

ANSWER: Hayden will get $\frac{1}{3}$ of the fruit bar.

Use the information to draw a whole and write a fraction.

1. Ayana and her 3 sisters are going to share a pizza equally. How much of the pizza will Ayana get?

2. Mrs. Jackson has 10 students in her music class. She is going to give each of the students an equal piece of the pie she made. How much of the pie will 3 students get?

3. Shauna has a granola bar that she divided into 6 equal pieces. She has eaten 4 of the pieces of the granola bar already. Show how many pieces of the granola bar Shauna has eaten.

4. Diego and his 7 friends are going to share a cake equally. How much of the cake will Diego get?

Fractions of a Set

Fractions can also represent groups, or sets, of items.

Josie has 6 teddy bears. Three of the teddy bears are white, and the other three teddy bears are gray.

EXAMPLE: Write a fraction to show how many white teddy bears Josie owns.

Step 1: Decide how many items are in the set.

Josie has 6 stuffed teddy bears.

Step 2: Write a fraction.

How many teddy bears are there in all? 6

Place this number in the denominator.

$$\frac{}{6}$$

Step 3: How many teddy bears are white? 3

Place this number in the numerator.

$$\frac{3}{6}$$

ANSWER: $\frac{3}{6}$ of Josie's teddy bears are white.

Use the information to draw a set and write a fraction.

1. Brittany has 8 heart barrettes and 2 circle barrettes. Write a fraction to show how many barrettes are heart.

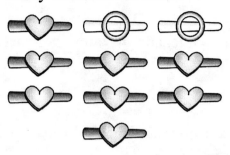

2. There are 5 racecars and 3 convertible cars in Jamar's car collection. Write a fraction to show how many convertible cars are in his collection.

3. Mr. Smith's dog had 3 puppies. Two of the puppies are spotted and one puppy has solid colored fur. Write a fraction to show how many puppies have spots.

4. There are 2 books on elephants on the shelf. There are 6 books in all on the shelf. Write a fraction to show how many of the books on the shelf are NOT about elephants.

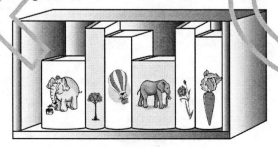

Fractions that Equal the Whole

At times, the numerator and the denominator will be the same number. This means the fraction equals the entire whole or the set.

EXAMPLE: Write a fraction that equals the entire set.

Step 1: Decide how many items are in the set.

There are 8 ducks at the pond. All of the ducks are swimming.

Step 2: Write a fraction.

How many ducks are there in all? 8

Place this number in the denominator.

$$\frac{}{8}$$

Step 3: How many ducks are swimming? 8

Place this number in the numerator.

$$\frac{8}{8} = 1$$

ANSWER: $\frac{8}{8}$ of the ducks at the pond are swimming.

Use the information to write a fraction.

1. There are a total of 10 cars. All of the 10 cars are blue. Write a fraction to show how many cars are blue.

2. There are 3 polar bears. All 3 polar bears are eating fish. Write a fraction to show how many polar bears are eating fish.

3. There are 6 markers in the box. All 6 of the markers are new. Write a fraction to show how many of the markers are new.

4. There are 8 monkeys at the zoo exhibit. All 8 of the monkeys are sleeping. Write a fraction to show how many of the monkeys are sleeping.

Comparing Fractions

In the section, "Fractions of a Whole," we divided a fruit bar equally into 3 pieces for Hayden to share with his two friends.

Suppose Hayden wants to have 2 pieces, but still wants everyone to have the same amount of equal pieces.

EXAMPLE: Find a way Hosea and his two friends can each have 2 equal pieces of the fruit bar.

Step 1: Decide how many pieces the fruit bar needs to be divided into.

There is Hosea and his two friends. This means there are 3 friends. Each friend wants 2 pieces. Multiply the amount of friends by the amount of pieces they want.

3(friends) × 2(pieces each friend wants) = 6

The fruit bars should be divided into 6 equal pieces

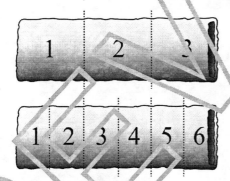

Step 2: Write a fraction.

How many pieces are there in all? 6

Place this number in the denominator.

$$\frac{\ }{6}$$

Step 3. How many pieces will Hosea get of the fruit bar? 2

Place this number in the numerator.

$$\frac{2}{6}$$

Notice how $\frac{1}{3}$ is the same amount as $\frac{2}{6}$.

This means the fractions are equal to each other.

ANSWER: Hosea will get $\frac{2}{6}$ of the fruit bar.

Decide if the models show equal amounts.

1.

4.

2.

5.

3.

6.

Chapter 4 Review

Use N for numerator and D for denominator to label the parts of the fractions.

1. $\dfrac{5}{6}$

2. $\dfrac{1}{3}$

3. $\dfrac{4}{10}$

Write a fraction to name the shaded pieces of the model.

4.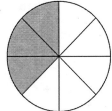

5.

Use the information to draw a whole and write a fraction.

6. Armando and his 5 friends are going to share a pizza equally. How much of the pizza will Armando get?

7. Kenna is sharing her banana with 2 friends. How much of the banana will Kenna get?

Use the information to draw a set and write a fraction.

8. Five of the singers in the choir are going to competition. There are 8 singers in total in the choir. Write a fraction to show how many singers will be going to competition.

9. There are 1 solar system bean bag and 1 truck bean bag. Write a fraction to show how many solar system bean bags are in the set.

Use the information to write a fraction.

10. There are a total of 10 bicycles. All of the 10 bicycles are red. Write a fraction to show how many bicycles are red.

11. There are 6 frogs. All 6 frogs are hopping. Write a fraction to show how many frogs are hopping.

Decide if the models show equal amounts.

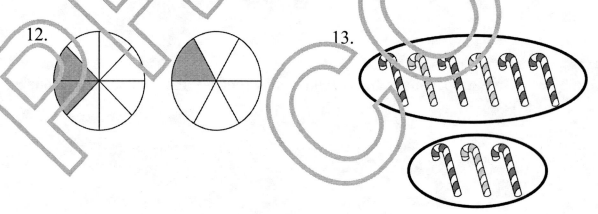

12.

13.

Chapter 4 Test

1. There are 8 kittens at the pet store. Five of the kittens are gray. The rest of the kittens are black. Which fraction shows how many of the kittens are black?

A $\dfrac{8}{8}$

B $\dfrac{5}{8}$

C $\dfrac{3}{8}$

2. What number is the numerator?

$$\dfrac{7}{8}$$

A 7
B 8
C 78

3. Which fraction represents the shaded part of the model?

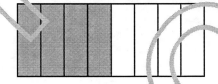

A $\dfrac{4}{8}$

B $\dfrac{8}{4}$

C $\dfrac{4}{12}$

4. Jake checked out 6 books from the library. Four of the books are about space and the rest of the books are about trains. Which fraction shows how many books are about trains?

A $\dfrac{4}{6}$

B $\dfrac{2}{6}$

C $\dfrac{6}{2}$

5. Which model shows an equal amount to the model below?

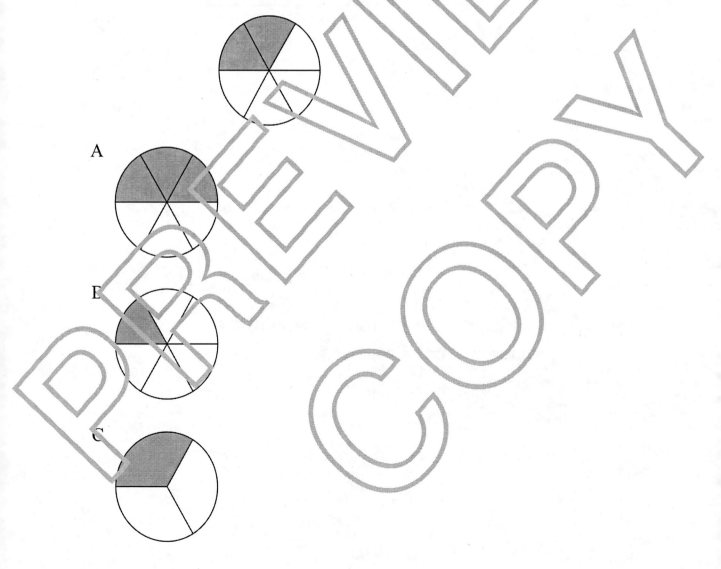

A

B

C

6. There are 3 sandwiches in the picnic basket. All 3 of the sandwiches are peanut butter and jelly. Which fraction shows how many sandwiches are peanut butter and jelly?

A $\dfrac{3}{1}$

B $\dfrac{3}{6}$

C $\dfrac{3}{3}$

7. Tiffany has 2 silver bracelets and 1 gold bracelet. Which fraction shows how many of the bracelets are silver?

A $\dfrac{1}{2}$

B $\dfrac{1}{3}$

C $\dfrac{2}{3}$

8. Gelisa and her 3 friends are going to share a pizza equally. How much pizza will Gelisa get?

A $\dfrac{1}{2}$

B $\dfrac{1}{3}$

C $\dfrac{1}{4}$

Chapter 5
Math Reasoning

This chapter covers the following Georgia Performance Standards:

| M2N | Number and Operations | M2N5a, b |

Using <, >, and = to Compare Amounts

You know that when you see + you add. There are also symbols you can use to compare. The symbol < means less than. The symbol > means more than. The symbol = means equal.

Theo wants to compare the number of toy trains he has to the number of toy helicopters he has.

EXAMPLE: Use a symbol to compare amounts.

Step 1: Decide how many items are in each group

6 toy trains

5 toy helicopters

Step 2: Decide which side has more.
There are more on the toy train side.

Step 3: Decide which symbol goes between the amounts.

6 toy trains   5 toy helicopters

There are more toy trains than toy helicopters. So, the more than symbol, >, should be used.

ANSWER: 6 toy trains > 5 toy helicopters
6 is **more than** 5

Use $>$, $<$, or $=$.

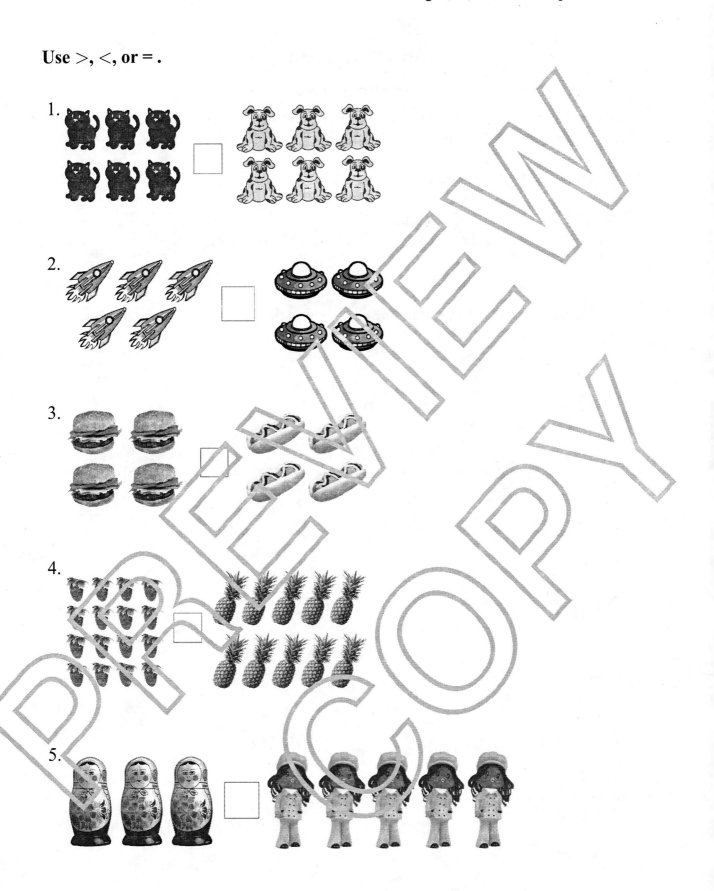

1.

2.

3.

4.

5.

Comparing Numbers

Comparing numbers is just like comparing amounts.

Martel wants to compare the number of laps he ran around the track to the number of laps Dylan ran around the track.

EXAMPLE: Use a symbol to compare amounts.

Step 1:   Decide which number is greater.
Martel ran 23 laps. Dylan ran 32 laps.

32 is greater than 23.

Step 2:   Write both numbers. Decide which symbol goes between the numbers.

23 $\boxed{<}$ 32

ANSWER: 23 < 32
23 is **less than** 32.

Use >, <, or = . One way to remember is the arrow always points to the lowest number.

1. 112 $\square$ 102

2. 24 $\square$ 42

3. 602 $\square$ 62

4. 701 $\square$ 710

5. 567 $\square$ 678

6. 905 $\square$ 905

7. 14 $\square$ 13

8. 136 $\square$ 136

9. 89 $\square$ 88

10. 123 $\square$ 321

Comparing Expressions

Amelia has 26 dollars. Her sister had 50 dollars but spent 25 dollars. Compare the money amounts.

EXAMPLE: Use a symbol to compare amounts.

Step 1: Write the information you need to compare.

26 ☐ 50 – 25

Step 2: Subtract.

26 ☐ 25

Step 3: Decide which symbol goes between the numbers.

26 ⟦ > ⟧ 25

ANSWER: 26 > 25
26 is **more than** 25.

Use >, <, or = .

1. 112 – 10 ☐ 102

2. 20 + 7 ☐ 50 – 27

3. 6 × 9 ☐ 9 × 6

4. 8 × 5 ☐ 7 + 34

5. 100 – 25 ☐ 6 × 9

6. 10 + 10 ☐ 4 × 5

7. 89 – 17 ☐ 13 + 54

8. 178 + 0 ☐ 178 × 0

9. 89 + 88 ☐ 88 + 89

10. 256 – 200 ☐ 200 + 256

Using +, −, and × to Find Missing Values

You need to use symbols to find the missing value in number sentences.

EXAMPLE: Use the symbol to find the missing value.

Step 1:   Decide what symbol is used.

$33 - \square = 11$

The subtraction symbol is being used.

Step 2:   Rewrite the number sentence. **Remember:** At times, you might need to use the opposite math operation to write a new number sentence.

$33 - \square = 11$

$33 - 11 = \square$

Step 3:   Decide what value subtracted from 33 will equal 11. The missing value is 22.

$33 - \boxed{22} = 11$

ANSWER: $33 - \boxed{22} = 11$

Find the missing value.

1. $300 + \square = 500$

2. $\square \times 5 = 35$

3. $175 - 50 = \square$

4. $\square + 89 = 172$

5. $8 \times 10 = \square$

6. $4 \times \square = 24$

7. $345 - \square = 217$

8. $99 + 101 = \square$

9. $\square - 57 = 129$

10. $782 + \square = 999$

Using +, –, and × in Problem Solving

You also need to use symbols to find the missing value in problem solving.

Zola has 27 stickers. Her mom gave her 34 more stickers. How many stickers does Zola have in all?

EXAMPLE: Use the symbol to find the missing value.

Step 1: Look at clue words in the problem to decide what symbol should be used.

The problem says Zola's mom gave her 34 **MORE** stickers. More is a clue word to add.

Step 2: Write a number sentence

$27 + 34 = \Box$

Step 3: Find the missing value. The missing value is 61.

$27 + 34 = \boxed{61}$

ANSWER: $27 + 34 = \boxed{61}$

Write a sentence and find the missing value.

1. Kasey is making 25 smoothies for her classmates. She has made 17. How many does she have left to make?

2. Mrs. Bernard cut 9 ribbons. Each ribbon is 7 inches long. What is the length of all the ribbons together?

3. Ross and his family are driving 234 miles to his grandmother's house. They have driven 198 miles. How many more miles do they need to drive?

4. Leela puts 4 kitten figures on each of the 3 shelves. How many kitten figures are on the shelves?

5. Darrius has collected 56 postage stamps. His uncle gave him 28 more postage stamps. How many postage stamps does Darrius have in all?

Chapter 5 Review

Use >, <, or = .

1.

2.

3.

4.   146 ☐ 86ა   9.   9 x 8 ☐ 8 x 9

5.   4ა ☐ 25   10.   98 ₋ 45 ☐ 64 + 34

6.   610 ☐ 622   11.   89 x 0 ☐ 89 + 0

7.   121 ☐ 102   12.   22 + 24 ☐ 24 + 22

8.   711 ☐ 711   13.   345 − 123 ☐ 100 + 150

Find the missing value.

14. $421 + \boxed{} = 896$

15. $\boxed{} \times 8 = 32$

16. $563 - 60 = \boxed{}$

17. $\boxed{} + 67 = 374$

Write a sentence and find the missing value.

18. Turell gave 4 friends 6 crackers each. How many crackers in all did Turell give his friends?

19. Virginia has 205 blocks. She use 175 blocks to build a tower. How many of the blocks did Virginia not use to build her tower?

20. John has 36 toy trains. His friend brings over 55 of his own toy trains. How many toy trains in all does John and his friend have?

Chapter 5 Test

1. Which symbol should be used to compare the numbers?

456 ☐ 654

A >
B <
C =
.

2. Cooper has 34 baseball cards. Bob has 43 baseball cards. Which comparison matches the information?

A 34 = 43
B 34 > 43
C 34 < 43

3. Which symbol should be used to compare the expressions?

8 x 8 ☐ 9 x 8

A >
B <
C =

4. Which symbol should be used to compare the numbers?

988 ☐ 989

A >
B <
C =

5. Mr. Duncan gave 3 pencils to each of the 10 students. How many pencils in all did he give to the students?

A 7
B 13
C 30

6. Leela ate 4 bites of her sandwich. Hannah ate 5 bites of her own sandwich. Which comparison matches the information?

   A $4 = 5$

   B $4 > 5$

   C $4 < 5$

7. Which symbol should be used to compare the expressions?

   $34 + 0 \square 34 \times 0$

   A $>$

   B $<$

   C $=$

8. Which symbol should be used to compare the amounts?

   A $>$

   B $<$

   C $=$

9. Lin has 6 more crackers than Zoe. Lin has 10 crackers. Which comparison matches the information?

   A $4 > 10 - 6$

   B $4 = 10 - 6$

   C $4 < 10 - 6$

10. Chris decided to practice his music for 50 minutes each week. This week, he has practiced 27 minutes. How many minutes of practice does he have left?

   A 23

   B 25

   C 77

11. Tyrice has 4 more toy trains than George. George has 14 toy trains. Which comparison matches the information?

   A $14 > 14 + 4$

   B $14 = 14 + 4$

   C $14 < 14 + 4$

12. What is the missing number?

   $6 \times \boxed{} = 30$

   A 4

   B 5

   C 6

13. Which symbol should be used to compare the expressions?

   $345 + 67 \ \boxed{} \ 700 - 300$

   A $>$

   B $<$

   C $=$

14. What is the missing number?

   $672 - \boxed{} - 168$

   A 162

   B 504

   C 516

15. Which symbol should be used to compare the amounts?

   A $>$

   B $<$

   C $=$

Chapter 6
Measurement

This chapter covers the following Georgia Performance Standards:

M2N	Number and Operations	M2N1c
M2D	Measurement	M2M1a, b, c
		M2M2
		M2M3

Length Measurement

There are two different types of length measurement. Standard measurement is mostly used in the United States. Some common standard units used are inch, foot, and yard.

Another type of length measurement is metric. Centimeters and meters are metric.

One important rule in measuring is to not mix up the units. If you start measuring in standard, do not change over to metric.

EXAMPLE: What is the length of this crayon to the nearest inch?

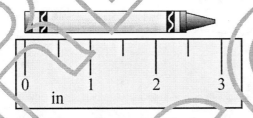

Step 1: Line up one side of the crayon on the zero mark of the ruler.

Step 2: Find the mark on the ruler that is closest to the other end of the crayon. In this drawing, the end of the crayon is closer to 3 inches than 2 inches.

Step 3: Write your answer, 3 inches.

ANSWER: 3 inches

Measure each item to the nearest inch.

1.

A

0		1		2	3

in

2.

0		1		2		3	4

in

3.

0		1		2		3		4	5

in

Measure to the Nearest Centimeter

Measuring centimeters is like measuring inches. The difference is you use a centimeter ruler.

EXAMPLE: How long is this key chain?

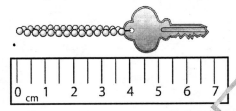

Step 1: Line up one side of the key chain on the zero mark of the ruler.

Step 2: Find the mark on the ruler that is closest to the other end of the key chain. In this drawing, the end of the key chain is closer to 7 centimeters than 6 centimeters.

Step 3: Write your answer, 7 centimeters.

ANSWER: 7 centimeters.

Measure each item to the nearest centimeter.

1.

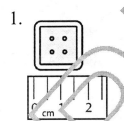

2.

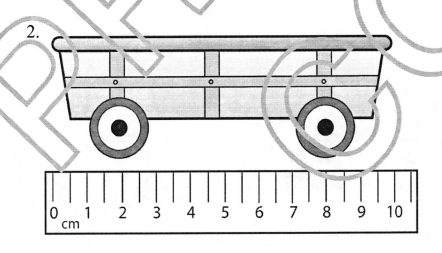

3.

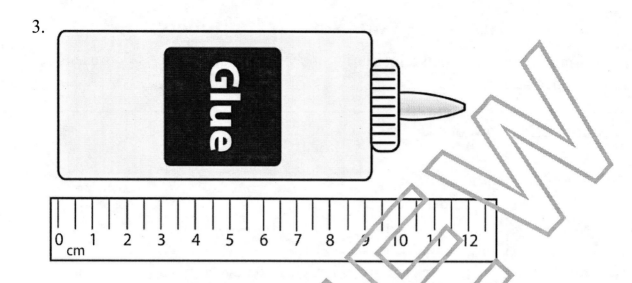

Comparing Lengths

Suppose you are putting items away on a shelf. You know the shelf will hold an item up to 8 inches or 20 centimeters long.

One way to make sure the item fits is to measure it by both length measurements.

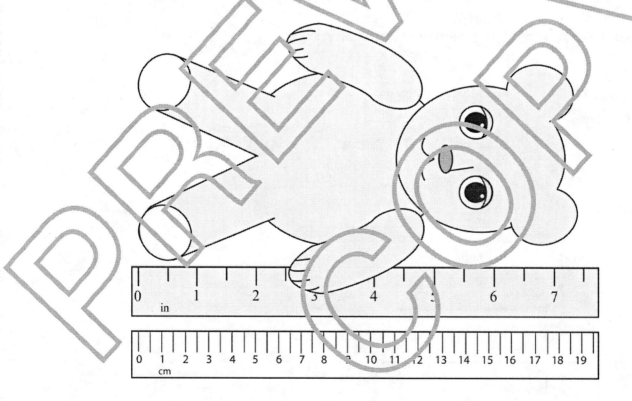

The bear measures less than 8 inches and less than 20 centimeters. It should fit on the shelf.

Tell whether the item would fit on an 8 inches or a 20 centimeters shelf.

1.

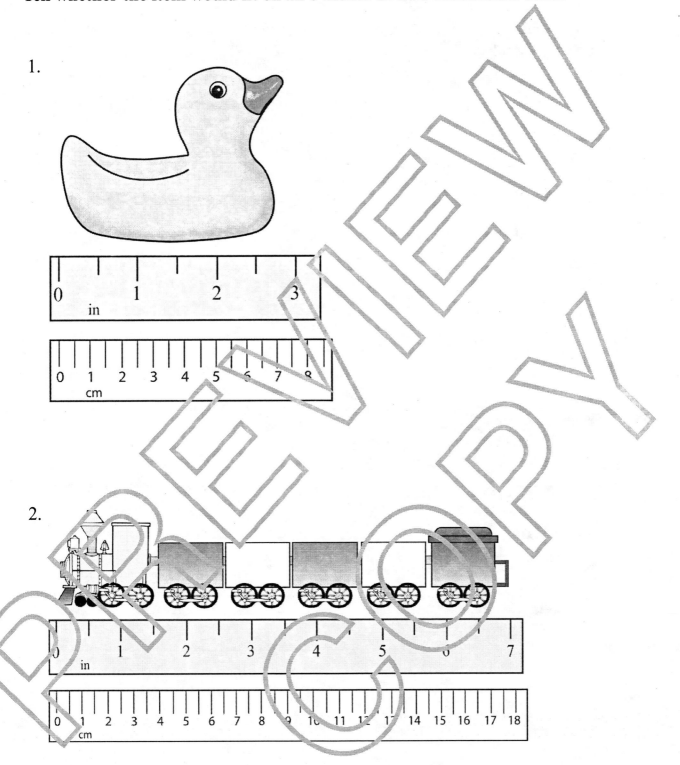

2.

Estimating Lengths

At times you might not need an exact measurement. Learning to estimate lengths can be very useful.

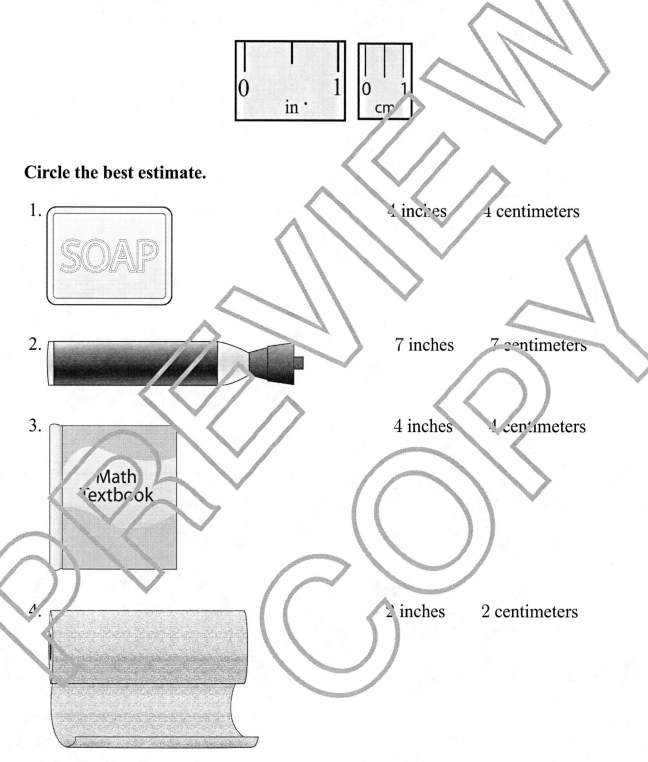

Circle the best estimate.

1.   4 inches   4 centimeters

2.   7 inches   7 centimeters

3.   4 inches   4 centimeters

4.   2 inches   2 centimeters

Using the Best Tool

Choosing the best tool or unit to use to measure is important.

EXAMPLE: Suppose you want to measure the length of a bicycle. Would you use centimeters or meters?

Step 1: Think about the length of a bicycle.

Step 2: Decide if the length of the bicycle is closer to the height of a chair or the width of you finger.

Step 3: A bicycle is closer to the height of a chair. So, you would most likely use meters to measure it.

You could use centimeters. However, it would take a much longer time to measure the bicycle.

Circle the best unit to use to measure each item.

Use the table to help you.

1 meter	=	100 centimeters
1 yard	=	3 feet
1 foot	=	12 inches

1. Adult Shoe   yard   inch

2. Dinner Fork   meter   centimeter

3. Notebook   inch   meter

4. Soccer Field   meter   centimeter

5. Penny   yard   centimeter

Estimating and Measuring Temperature

Knowing how to estimate and measure temperature can be useful. It can help you decide what clothes to wear.

EXAMPLE: Estimate the temperature based on the clothes the child is wearing.

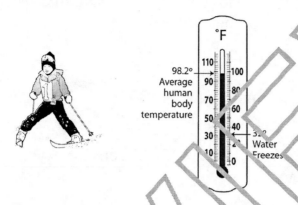

Step 1:   Look at the clothes.

Step 2:   Decide if the clothes would be worn when the temperature is hot or cold.

Step 3:   Use the thermometer as a guide.

Step 4:   Record your estimate, 25°.

Estimate the temperature.

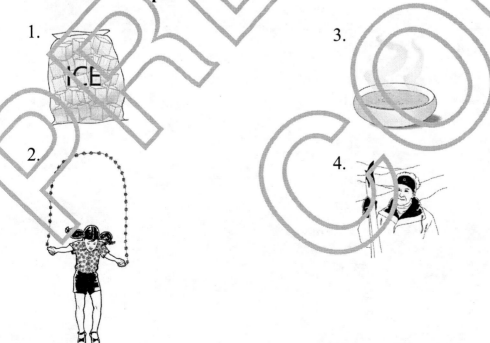

1.

2.

3.

4.

EXAMPLE: Now measure the temperature.

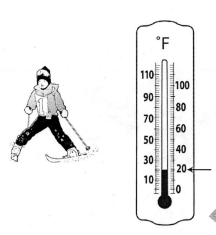

Step 1: Find the top of the shaded part of the thermometer.

Step 2: Look at the mark on the thermometer.

Step 3: This thermometer shows 20°.

Step 4: Record your measurement, 20°.

Was the estimate reasonable?

ANSWER: Yes, because 25° is close to 20°.

Find the temperature.

1.
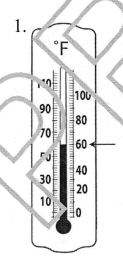

2.

3.
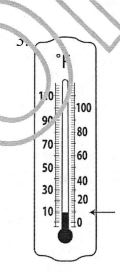

Measuring Time

Below is a digital clock.

Analog clocks use hands. The short hand shows the hour. The long hand shows the minute.

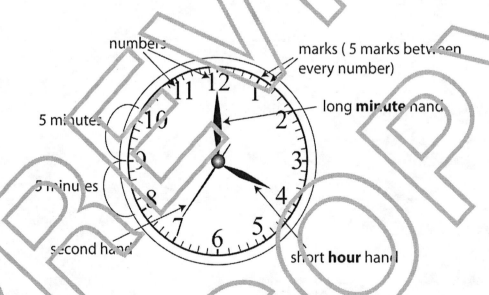

There are 60 minutes in one hour. There are 24 hours in one day.

1 hour = 60 minutes

1 day = 24 hours

EXAMPLE: What time is it?

Step 1: Look at the short hand on the clock. This is the hour hand.

Step 2: The hour hand is pointing to 2.

Step 3: Now look at the long hand. This is the minute hand.

Step 4: The minute hand is pointing to 12. When the minute hand points to 12, it tells us it is the exact hour.

ANSWER: It is 2:00, or 2 o'clock.

EXAMPLE: What time is it?

Step 1: Look at the short hand on the clock. It is between the 4 and 5.

Step 2: When the hour hand is in between numbers, use the hour that is less, 4.

Step 3: Look at the minute hand. It is on the 6. Each number on the clock stands for 5 minutes when using the minute hand.

Step 4: Start at 1 and skip count by 5's until you reach the 6.

5, 10, 15, 20, 25, 30

ANSWER: The time is 4:30.

Tell the time.

1.

2.

3.

Tell what number the minute hand should be on an analog clock.

4.

5.

6.

Counting Money

We use money to buy items or services.

Decimals are used to show how many dollars and how many cents.

The dollar sign, $, and the cents sign, ¢, are also used when using money.

The table shows some commonly used coins.

Penny	Nickel	Dime	Quarter
1 cent (1¢) $0.01	5 cents (5¢) $0.05	10 cents (10¢) $0.10	25 cents (25¢) $0.25

EXAMPLE: Austin used one dollar bill to buy a pen. The pen costs 75¢. How much change should he receive back?

Step 1: Change the dollar bill to a number value using a decimal. $1.00

Step 2: Change the item amount to a number value using a decimal. $0.75

Step 3: Decide on what operation to use. Since we are finding change, use subtraction.

Step 4: Use the operation.

$$\begin{array}{r} \$1.00 \\ -\$0.75 \\ \hline \$0.25 \end{array}$$

Step 5: Record your answer, $0.25.

ANSWER: $0.25

Use decimals to name the money shown.

1.

Quarter   Quarter   Quarter   Nickel   Nickel

2.

Quarter   Quarter   Quarter   Dime

Quarter   Quarter   Quarter   Penny

3.

$5.00   Quarter   Quarter

4.

$1.00

Copyright © American Book Company

Use decimals to find the change.

5. Lillian buys a toy cat for $4.25. She hands the clerk a $5.00 bill. What should her change be?

6. Miles wants to buy a comic book for $6.50. He ha⁢

Does he have enough money to buy the comic book? Explain.

7. Bailey gives the clerk

for an $8.25 art set. How much change should she get back?

Chapter 6 Review

1. What is the length of the crayon to the nearest inch?

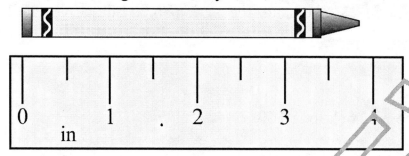

2. What is the length of the crayon to the nearest centimeter?

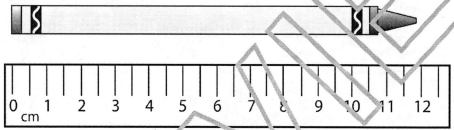

Use decimals to find the change.

3. Samantha wants to buy a drink and a snack for $3.70. She hands the clerk a five dollar bill. How much change should she get back?

Circle the best estimate.

4.

Banana

20 inches   20 centimeters

5.

Stapler

8 inches   8 centimeters

Circle the best unit to use to measure each value.

6. Flag pole   yard   inch

7. Cell phone   meter   centimeters

Estimate the temperature.

8.

9.

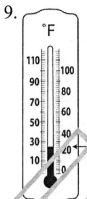

Find the time.

10.

11.

Name the number the minute hand should be on.

12.

13.

Use decimals to name the money shown.

14.

$1.00

Quarter   Quarter   Quarter

Dime   Dime   Penny   Penny   Penny   Penny

Dime   Dime   Penny   Penny   Penny   Penny

Nickel   Nickel

Chapter 6 Test

1. What is the length to the nearest inch?

A 1 inch
B 2 inches
C 3 inches

2. What is the length to the nearest centimeter?

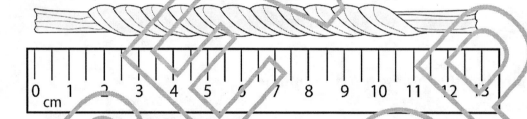

A 10 centimeters
B 11 centimeters
C 13 centimeters

3. Which length listed is the best estimate?

A 20 inches
B 20 centimeters
C 20 meters

4. Which of these units would you **NOT** use in standard measurement?

    A inch

    B foot

    C meter

5. Which length listed is the best estimate?

    A 5 inches

    B 5 feet

    C 5 yards

6. Which temperature listed is the best estimate?

    A 10 degrees

    B 30 degrees

    C 60 degrees

7. Which temperature listed is the best estimate?

    A 10 degrees

    B 25 degrees

    C 80 degrees

8. What is the time?

A 3:30

B 5:30

C 4:30

9. What is the temperature?

A 75 degrees

B 85 degrees

C 90 degrees

10. What is the time?

A 1:05

B 1:25

C 2:25

11. What number should the minute hand be on?

   A 3
   B 5
   C 6

12. How much money does Bobbi have?

   A $0.90
   B $1.00
   C $1.20

13. How much money does Adalia have?

$5.00

   A $1.65
   B $1.75
   C $5.65

14. Hannah gives the clerk a one dollar bill to buy a drink for $0.85. How much change should she receive?

   A $0.15
   B $0.85
   C $1.00

Chapter 7
Plane Geometry

This chapter covers the following Georgia Performance Standards:

M2G	Geometry	M2G1

Closed and Open Shapes

Plane shapes are flat and have lines. The lines can be straight or curved.

Plane shapes can be open or closed.

EXAMPLE: Is this plane shape open or closed?

Step 1: Place your finger at the *A*.

Step 2: Move your finger along the lines of the shape until you get to the end of the line or where you first started.

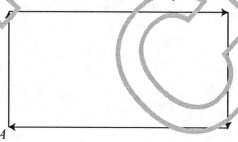

Step 3: You started at the bottom left corner of the shape and ended at the same spot. This is a closed plane shape.

ANSWER: closed

Tell whether each plane shape is open or closed.

1.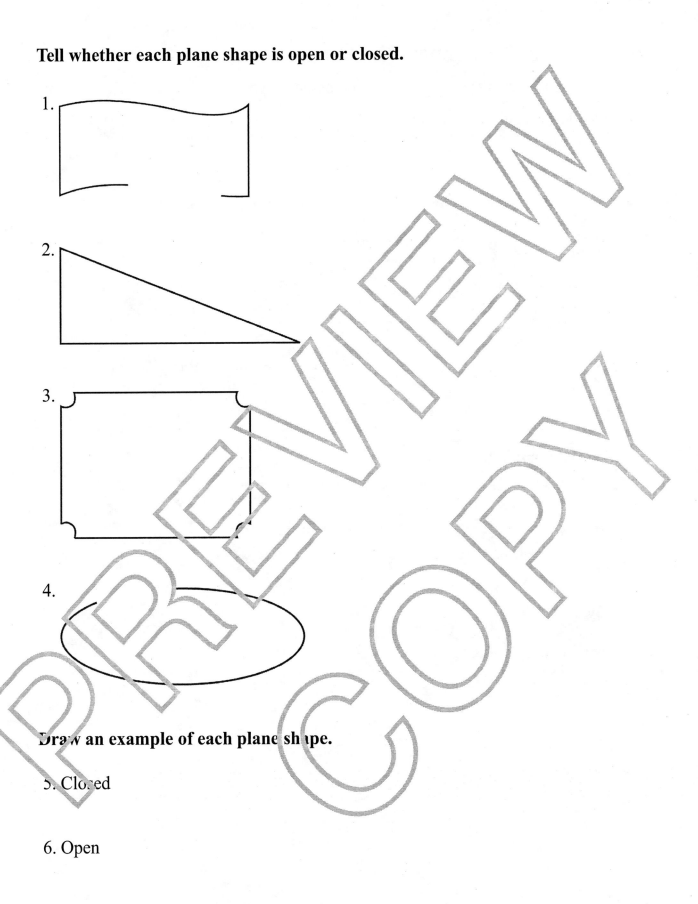

2.

3.

4.

Draw an example of each plane shape.

5. Closed

6. Open

Points, Lines, Line Segments, and Rays

Here are some terms to help you describe plane shapes.

A **point** is a specific location or position.

A **line** is straight and goes in opposite directions without ending.

A **line segment** is straight and has two endpoints. A line segment is part of a line.

A **ray** is straight and has one endpoint and the other side the line goes on forever.

Match the term to the item.

1. line

2. line segment

3. point

4. ray

A.

B.

C.

D.

Draw an example of each term.

5. Ray

6. Line

7. Point

8. Line Segment

Edges and Vertices

An **edge** is a line segment on a plane shape. Edges can also be called sides on a plane shape.

A **vertex** is a shared endpoint. More than one vertex are called vertices.

Notice the edges and vertices on this plane shape.

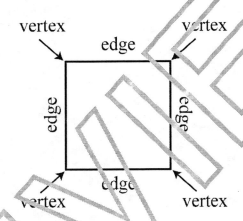

EXAMPLE: Locate the vertices.

Step 1: Look at the shape. Find the endpoints.

Step 2: Place an X on each endpoint.

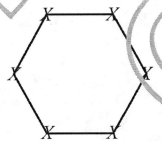

There are six vertices.

Place an X on each edge and circle each vertex.

1.

4.

2.

5.

3.

6.

Angles

You have learned some of the terms that will help you identify different plane shapes.

Another important term to know is angle.

An **angle** is two rays that come together at a vertex.

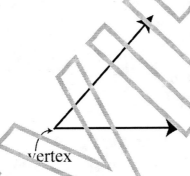

vertex

Angles are measured in degrees.

There are three main types of angles.

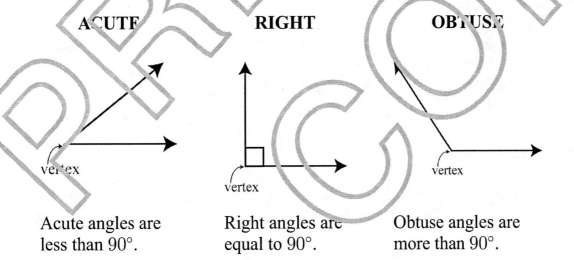

ACUTE	RIGHT	OBTUSE

vertex

vertex

vertex

Acute angles are less than 90°.

Right angles are equal to 90°.

Obtuse angles are more than 90°.

EXAMPLE: Identify the angle.

Step 1:   Locate the angle.

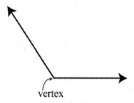

Step 2:   Look at the examples of acute, right, and obtuse angles. Does this angle look like a right angle? No

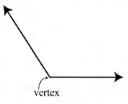

Step 3:   We know this is not a right angle. Does the angle look smaller or larger than a right angle? Larger

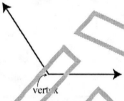

ANSWER: obtuse angle

Draw an example.

1. Right   2. Obtuse   3. Acute

Label each angle.

4.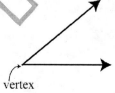

5.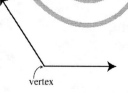

6.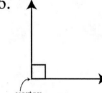

Classifying Plane Shapes

You have learned the terms to describe plane shapes. Now you are ready to use the terms to classify plane shapes.

Closed plane shapes can be classified by the number of edges, vertices, and angles.

Shape	Name	Edges	Vertices	Angles
△	Triangle	3	3	3
□	Square	4 all equal	4	4 all right angles
▭	Rectangle	4	4	4 all right angles
⏢	Trapezoid	4	4	4
◇	Quadrilateral	4	4	4
⬠	Pentagon	5	5	5
⬡	Hexagon	6	6	6

Fill in the chart by counting the number of edges, vertices, and angles.

Shape	Edges	Vertices	Angles
1.			
2.			
3.			
4.			
5.			

Draw a closed plane shape according to the given edges, vertices, and angles.

6. 3 edges, 3 vertices, 3 angles

8. 4 edges, 4 vertices, 4 angles

7. 5 edges, 5 vertices, 5 angles

9. 6 edges, 6 vertices, 6 angles

Using Plane Shapes to Make More Plane Shapes

Some plane shapes can be divided into pieces to make new plane shapes. Plane shapes can also be combined to make new shapes.

EXAMPLE: Identify the new plane shapes.

Step 1:   Identify the plane shape.

This is a rectangle.

Step 2:   Draw a line down the middle to divide the rectangle in half.

Step 3:   Identify the new plane shapes.

ANSWER: When the rectangle was divided, it became 2 squares.

EXAMPLE: Identify the new plane shapes.

What if the same rectangle from the previous example were divided another way?

Step 1:   Identify the plane shape.

This is a rectangle.

Step 2:   Draw a line from corner to corner to divide the rectangle in half.

Step 3:   Identify the new plane shapes.

ANSWER: When the rectangle was divided, it became 2 triangles.

Draw a line on the plane shape to make new plane shapes. Then name each new plane shape.

1.

2.

3.

4.

EXAMPLE: Identify the new plane shape.

Plane shapes can also be combined to make a new plane shape.

Step 1: Identify the plane shapes.

Here are two rectangles.

Step 2: Combine the plane shapes.

Step 3: Identify the new plane shape.

ANSWER: When the two rectangles were combined, they became a square.

Combine the two plane shapes, and draw the new plane shape. Then name each new plane shape.

5.

6.

7.

8.

Chapter 7 Review

1. Draw a closed plane shape.

2. Draw an open plane shape.

.

Match the term to the definition.

3. line   A. straight with points on each end

4. line segment   B. straight with one end point and one side that goes on forever

5. point   C. specific location

6. ray   D. straight and does not end on opposite sides

Circle all the vertices and place X on all the edges.

7.

Label each angle.

8.

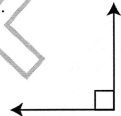

9.

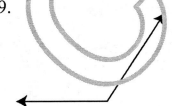

10.

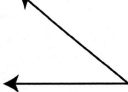

Combine the plane shapes to make a new plane shape.

11.

12.

13.

Fill in the chart below by counting the number of edges, vertices, and angles.

	Shape	Edges	Vertices	Angles
14.	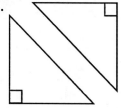			
15.				
16.				
17.				
18.				

Chapter 7 Test

1. Which is a closed shape?

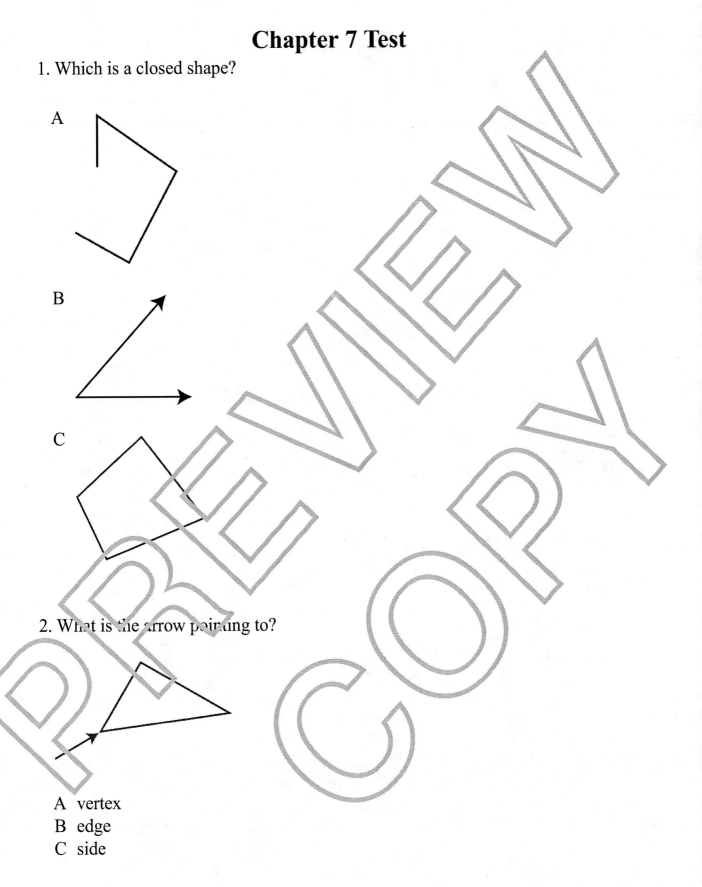

A

B

C

2. What is the arrow pointing to?

A vertex
B edge
C side

3. How many edges does a trapezoid have?

A 3
B 4
C 5

4. What type of angle is shown?

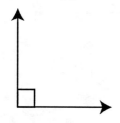

A right
B obtuse
C acute

5. How many end points does a ray have?

A 0
B 1
C 2

6. Which plane shape listed has exactly 3 angles?

A hexagon
B rectangle
C triangle

7. How many end points does a line segment have?

A 2
B 1
C 0

8. How many vertices does this plane shape have?

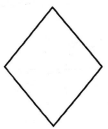

  A 2
  B 3
  C 4

9. What type of angle measures more than 90°?

  A obtuse
  B right
  C acute

10. Which plane shape listed has 4 edges?

  A trapezoid
  B pentagon
  C hexagon

11. Which plane shape has 4 right angles?

  A pentagon
  B square
  C trapezoid

12. Which new plane shape can be made with these two plane shapes?

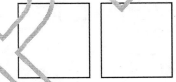

  A circle
  B rectangle
  C square

Chapter 8
Solid Geometry

This chapter covers the following Georgia Performance Standards:

| M2G | Geometry | M2G2a, b |
| | | M2G3 |

Naming Solid Shapes

Solid shapes have length, width, and height. They are not flat like a piece of paper. Solids are three-dimensional in shape.

Here are some common solid shapes.

Rectangular Prism	Cylinder	Cone
Sphere	Cube	Pyramid

Name each solid.

1.

2.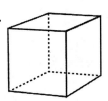

3.

4.

5.

6.

Name the solid that is similar to each item.

7.

8.

9.

10.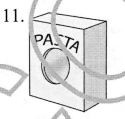

11.

12.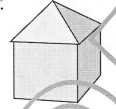

Sorting Solid Shapes

One way to sort solid shapes is to find out if the shape has flat surfaces or curved surfaces. Some solids have both.

You can also sort solid shapes by determining if the solid can slide, roll, or do both.

Solids with flat surfaces can slide. Solids with curved surfaces can roll. If a solid has both flat and curved surfaces it can slide and roll.

Solids with flat surfaces	Solids with curved surfaces	Solids with both flat and curved surfaces

EXAMPLE: Identify if the solid can slide, roll, or do both.

Step 1:   What type of surfaces does the solid have?

Step 2:   The solid has both flat and curved surfaces. A flat surface can slide. A curved surface can roll.

Step 3:   Since the solid has both flat and curved surfaces, this solid can slide and roll.

ANSWER: both slide and roll

Decide if the solid has flat, curved, or both type of surfaces.

1.

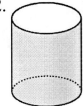

3.

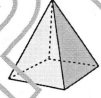

2.

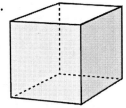

4

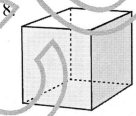

Decide if the solid can slide, roll, or do both.

5.

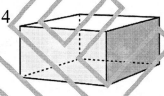

7.

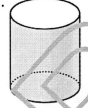

6.

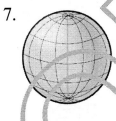

8.

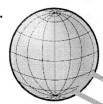

Identifying Angles

Another way to sort solids is to identify the angles in the solid.

There are three main types of angles.

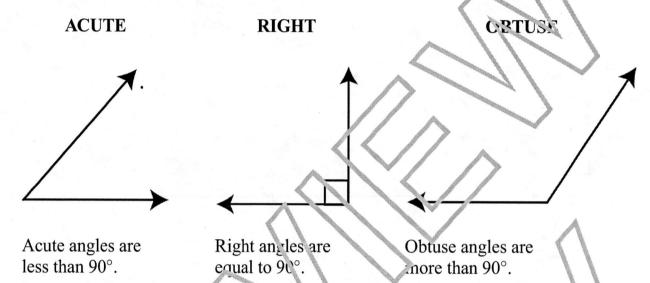

| **ACUTE** | **RIGHT** | **OBTUSE** |

Acute angles are less than 90°.

Right angles are equal to 90°.

Obtuse angles are more than 90°.

EXAMPLE: Identify the angle on the solid.

Step 1: Locate the angle.

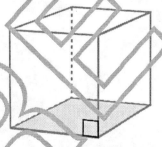

Step 2: Look at the examples of acute, right, and obtuse angles. Does this angle look like a right angle? Yes

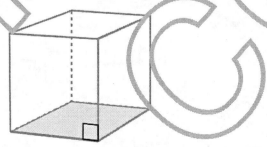

Step 3: This angle is the same size as a right angle. So, it is a right angle.

ANSWER: right angle

Identify the angle on the solid.

1.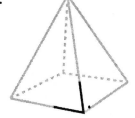

2.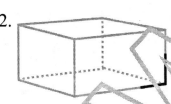

Finding Faces on Solids

A **face** is the plane shape, or flat surface, found on a solid.

EXAMPLE: Identify the faces on the solid.

Step 1: Trace each flat side of the solid.

Step 2: Look at the faces or plane shapes. Name each face.

Circle, Circle

Step 3: The solid cylinder has two faces. Both of the faces on the cylinder are circles.

ANSWER: Two circles are on a cylinder.

Identify the types of faces on the solid.

1.
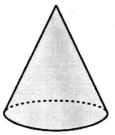

2.

3.
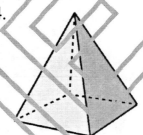

4.
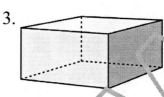

Circle the solid(s) on the right that have the plane shape on the left as one of its faces.

5.
  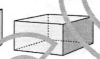

6.
  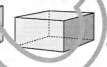

7.

8.

Classifying Solid Shapes

You have learned some ways to sort solid shapes. Now you are ready to classify solid shapes.

Solid shapes can be classified by the number of edges, vertices, and faces.

Shape	Name	Edges	Vertices	Faces
	Square Pyramid	8	5	□ △ △ △ △
	Cylinder	0	0	○ ○
	Cone	0	0	○
	Sphere	0	0	None
	Cube	12	8	□ □ □ □ □ □
	Rectangular Prism	12	8	▭ ▭ ▭ ▭ □ □

Fill in the chart below by counting the number of edges and vertices for each shape

	Shape	Edges	Vertices
1.	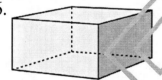		
2.			
3.			
4.			
5.			

Draw all the faces for each solid.

6.

9.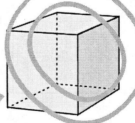

7.

10.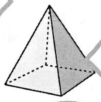

8.

11.

Using Solids to Make More Solids

Some solids can be divided into parts to make new solids. Solids can also be combined to make new solids.

EXAMPLE: Identify the new solids.

Step 1: Identify the solid.

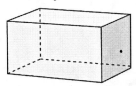

This is a rectangular prism.

Step 2: Draw a line down the middle to divide the rectangular prism in half.

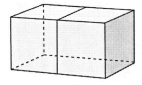

Step 3: Identify the new solids

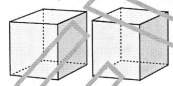

ANSWER: When the rectangular prism was divided, it became 2 cubes.

Draw lines on the solid to make new solids. Then name each new solid.

1.

2.

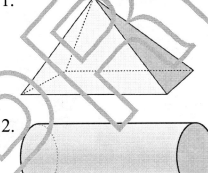

3.

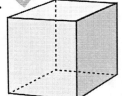

EXAMPLE: Identify the new solid.

Solids can also be combined to make solids.

Step 1:   Identify the solids.

Here are two cylinders.

Step 2:   Combine the solids.

Step 3:   Identify the new solid.

ANSWER: When the two cylinders were combined, they became a longer cylinder.
cylinder.

Combine the two solids. Then name each new solid.

4.
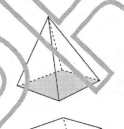

5.
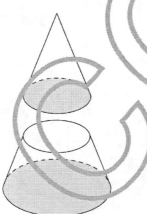

6.
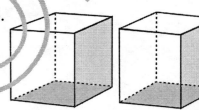

Chapter 8 Review

Name the solid.

1.

2.

3.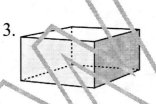

Name the solid that is similar to each object.

4.

6.

5.

7.

Name two solid shapes that can roll.

8. _____

9. _____

Name two solid shapes that can slide.

10. _____

11. _____

Name two solid shapes that can roll and slide.

12. _____

13. _____

Identify a right angle on the solid.

14.

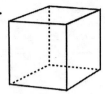

Name the faces on each solid.

15.

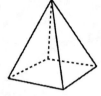

16.

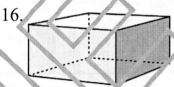

Fill in the table.

		Edges	Vertices
17.	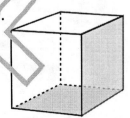		
18.			
19.			

Divide the solid into parts to make solids.

20.

21.

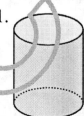

Chapter 8 Test

1. Which is similar to a cylinder?

A

B

C

2. Which solid has 12 edges?

A   rectangular prism
B   cylinder
C   cone

3. How many vertices does this solid have?

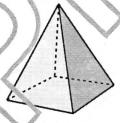

A   1
B   4
C   5

4. Which solid has only right angles?

A cone
B cube
C pyramid

5. How many triangle faces are on a square pyramid?

A 2
B 3
C 4

6. What is the name of this solid?

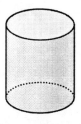

A cone
B cube
C cylinder

7. Which solid has a curved surface and no flat surfaces?

A sphere
B pyramid
C cylinder

8. How many edges does this solid have?

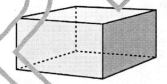

A 12
B 8
C 6

9. Which new solid can be made with these two solids?

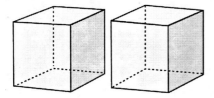

   A cube
   B rectangular prism
   C pyramid

10. How many square faces are on a cube?

   A 3
   B 4
   C 6

11. How many edges does a sphere have?

   A 0
   B 1
   C 5

12. Which solid has this plane shape as a face?

   A cone
   B square pyramid
   C rectangular prism

Chapter 9
Graphs

This chapter covers the following Georgia Performance Standards:

M2D	Data Analysis and Probability	M2D1a, b

Charts and Tables

There are many ways to organize and sort items. One way is to use a chart or table.

EXAMPLE: The table below lists the number of moons, stars, and suns in a picture. Find the number of stars in the picture.

Step 1: Look at the table. Find the word Stars under the Shape column.

Shape	Number
Moon	4
Stars	5
Suns	7

Step 2: Now look at the number next to stars. What number is next to stars? 5

ANSWER: There are 5 stars in a picture.

Use the table to answer the questions.

The table shows favorite ice-pop flavors.

Flavor	Number
Cherry	10
Grape	7
Lemon	4
Orange	7
Strawberry	5

1. Which is the favorite flavor?

2. How many people like orange as a favorite flavor?

3. How many people like strawberry as a favorite flavor?

4. How many people like the least favorite flavor?

5. How many people in total like lemon and grape as favorite flavors?

6. Which two flavors have the same number of votes?

Displaying information in charts or tables is a way to organize information.

EXAMPLE: Use the information to fill in the table.

Sandwich	Number
Ham	ⵚ I
Turkey	ⵚ III
Peanut Butter	III
Cheese	ⵚ

Step 1:   Look at the information. What types of sandwiches are in the lunch count? Write the type of sandwiches in the left column of a table. This is the lunch count for Mrs. Jackson's class.

Step 2:   Now look at the number of tallies for each sandwich. Fill in the right column with the sandwich totals.

Sandwich	Number
Ham	6
Turkey	8
Peanut Butter	3
Cheese	5

Use the information given to fill in the table.

This is the lunch count for side items.

Side Dish	Number
Carrot sticks	ⵚ
Salad	II
Fruit Cup	ⵚ IIII
Crackers	ⵚ I

Side Dish	Number
7.	11.
8.	12.
9.	13.
10.	14.

Venn Diagrams

Another way you can sort items is by using a Venn diagram. **Venn diagrams** group similar items by using overlapping circles.

EXAMPLE: Use the Venn diagram to find out how many black triangles there are.

Step 1: Look at the Venn diagram. Decide what is common in the left circle. All the items in the left circle are triangles.

Step 2: Now look at the right circle. Decide what is common in the right circle.
All the items in the right circle are black shapes.

Step 3: We know the items in the left circle are triangles, and the items in the right circle are black shapes. The items in the middle will have both the left and right circles in common. So, the middle area will have all the black triangles.

Step 4: Count the number of shapes in the middle that are triangles and black in color.

ANSWER: There are 3 black triangles.

Use the Venn diagram to answer the questions.

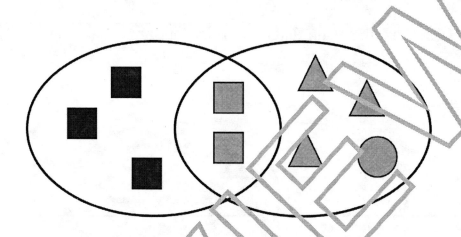

1. How many black squares are there?

2. How many gray shapes are there in all?

3. How many gray squares are there?

4. How many total gray and black shapes are there?

EXAMPLE: Use the information to fill in the Venn diagram.

Step 1: Look at the information. What categories are shown?

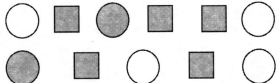

There are gray circles, white circles, and gray squares.

Step 2: Decide which items you will put in the left circle.

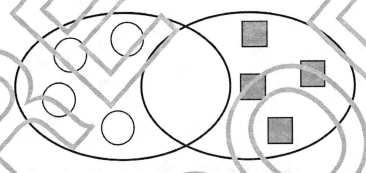

Step 3: Use the right side to put in items that are the opposite of what is on the left side of your Venn diagram.

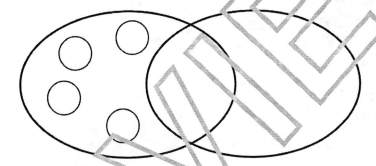

Step 4: Use the middle to show items that have characteristics of both the left and right sides.

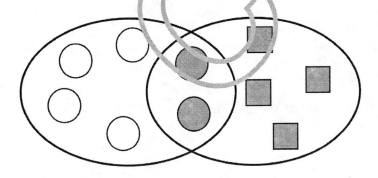

Use the information given to fill in the Venn diagram.

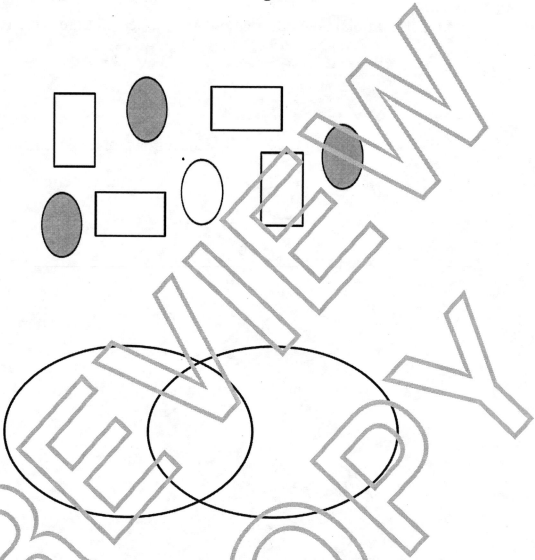

5. Draw a Venn diagram where the white oval should go.

6. Draw on the Venn diagram where the gray ovals should go.

7. Draw on the Venn diagram where the white rectangles should go.

Picture Graphs

Graphs are also used to sort and classify items into groups. **Picture graphs** use pictures or symbols to show the items being counted. Most picture graphs have a key that shows how much each picture is worth.

EXAMPLE: Find out how many children like banana smoothies.

Step 1: Look at the picture graph. Find the row with bananas. Then count how many bananas are listed.

Smoothies We Like					
Banana	🍌	🍌	🍌		
Grape	🍇	🍇			
Strawberry	🍓	🍓	🍓	🍓	

Key: Each picture = 2 children

There are 3 bananas listed.

Step 2: Now look at the key. How much is each picture worth? Each picture is worth 2.

Step 3: How many children in all like banana smoothies?
6

ANSWER: There are 6 children in all that like banana smoothies.

Use the picture graph to answer the questions.

Lunches We Like					
Hamburger	🍔	🍔	🍔		
Pizza	🍕	🍕			
Tacos	🌮	🌮	🌮	🌮	

Key: Each picture = 3 children

1. Which is the favorite lunch?

2. How many children like hamburger as a favorite lunch?

3. How many children like the least favorite lunch?

4. How many children in total like pizza and tacos as favorite lunches?

Use the information to fill in the picture graph.

Austin did a survey of what type of pets children in his class had at home. Each picture is equal to 2 children. Six classmates have dogs, four classmates have cats, and two classmates have fish.

Pets We Have					
Cat					
Dog					
Fish					

Key: Each picture = _____ children

5. Fill in the key.

7. Draw the amount for fish as pets.

6. Draw the amount for dogs as pets.

8. Draw the amount for cats as pets.

Bar Graphs

Bar graphs are a common way to show how items are organized and classified. Bars are used on the graph to show how many items each group contains.

EXAMPLE: Find out how many children like to swim at the pool.

Step 1: Look at the bar graph. Find the bar with pool.

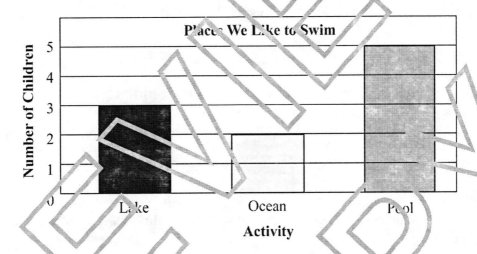

Step 2: Move your finger up the pool bar to the top of the bar. Then, move your finger to the left to see what number the bar stops at.

Step 3: How many children like to swim at the pool?
5

ANSWER: There are 5 children in all that like to swim at the pool.

Use the bar graph to answer the questions.

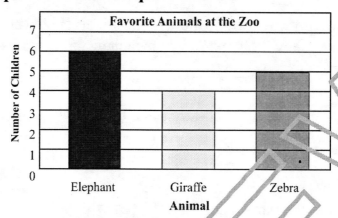

1. Which is the favorite animal to see at the zoo?

2. How many children like the giraffe as a favorite animal?

3. How many children like the second favorite animal?

4. How many children in total like elephants and zebras as favorite animals?

Use the information to fill in the bar graph.

Lillie did a survey of favorite sport to play. There are 5 classmates that like to play soccer the most. Golf is a favorite sport for 3 classmates, and 7 classmates like to bowl the most.

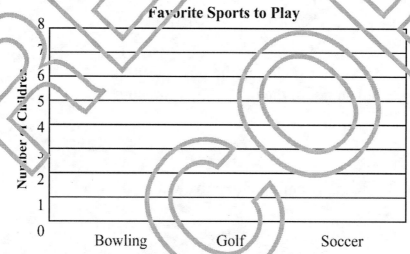

5. Draw a bar to show how many children like to bowl the most.

6. Draw a bar to show how many children like to play golf the most.

7. Draw a bar to show how many children like to play soccer the most.

Chapter 9 Review

Use the table to answer the questions.

Season	Number
Fall	5
Spring	8
Summer	12
Winter	7

1. Which is the favorite season?

2. How many people like winter as a favorite season?

3. How many people like the least favorite season?

4. How many people in total like fall and spring as favorite seasons?

Use the Venn diagram to answer the questions.

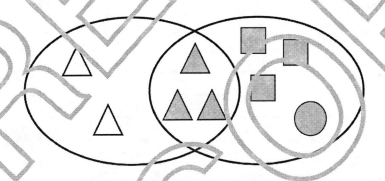

5. How many white triangles are there?

6. How many gray shapes are there in all?

7. How many gray triangles are there?

Use the information to fill in the picture graph.

Isabella did a survey of favorite vegetables. Each picture is equal to 2 children. Eight classmates said carrots were their favorite vegetable. Four classmates chose green beans, and two classmates picked broccoli.

Vegetables We Like					
Broccoli					
Carrots					
Green Beans					

Key: Each picture = _____ children

8. Fill in the Key.

9. Draw the amount for broccoli as a favorite vegetable.

10. Draw the amount for carrots as a favorite vegetable.

11. Draw the amount for green beans as a favorite vegetable.

Use the information to fill in the bar graph

Lennon did a survey of favorite colors. There are 8 classmates that like the color blue the most. Orange is a favorite color for 6 classmates and 5 classmates like the color green the most.

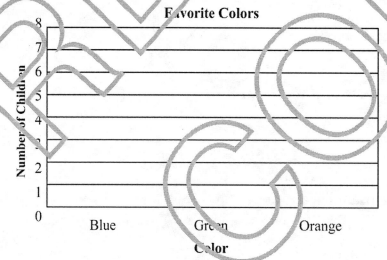

12. Draw a bar to show how many children like the color blue the most.

13. Draw a bar to show how many children like the color green the most.

14. Draw a bar to show how many children like the color orange the most.

Chapter 9 Test

Use the table to answer questions 1–4.

Month	Number
May	2
June	6
July	4
August	9

1. Which month had the most rainy days?

    A   August

    B   July

    C   June

2. How many days in total were rainy in July and June?

    A   4

    B   6

    C   10

3. How many days were rainy in May?

    A   2

    B   4

    C   6

4. How many more days were rainy in August than in June?

    A   3

    B   5

    C   6

Use the Venn diagram to answer questions 5–7.

5. How many black shapes are there in all?

   A 5

   B 8

   C 12

6. How many total shapes are there?

   A 5

   B 7

   C 12

7. How many triangles are there?

   A 4

   B 5

   C 7

Use the picture graph to answer questions 8–11.

Smoothies We Like					
Banana	🍌	🍌	🍌		
Grape	🍇	🍇			
Strawberry	🍓	🍓	🍓	🍓	

Key: Each picture = 2 children

8. How many children like banana as a favorite smoothie flavor?

    A 2

    B 4

    C 6

9. Which is the favorite smoothie flavor?

    A banana

    B grape

    C strawberry

10. How many children in total like banana and grape as favorite smoothies?

    A 10

    B 5

    C 3

11. How many children like the least favorite smoothie?

    A 2

    B 4

    C 6

Use the picture graph to answer questions 12–15.

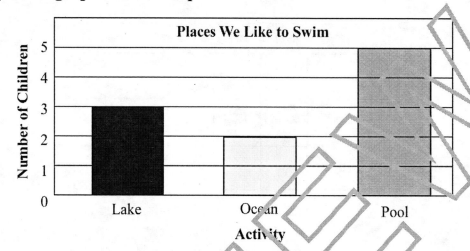

12. Which is the favorite place to swim?

    A lake

    B ocean

    C pool

13. How many more children like to swim at the pool than the ocean?

    A 3

    B 2

    C 1

14. How many children like to swim at the lake?

    A 2

    B 3

    C 5

15. How many children in total like to swim at the ocean and the lake?

    A 3

    B 5

    C 7

Practice Test 1

Part 1

1. What number matches $500 + 40 + 3$?

    A $500, 403$

    B 543

    C $5, 043$

M2N1a

2. Leo counted 104 cars on the train. Which figure models the number 104?

    A

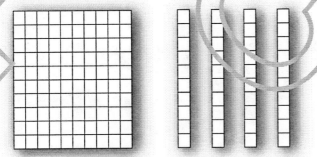

    B

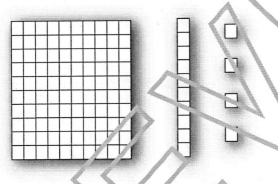

    C

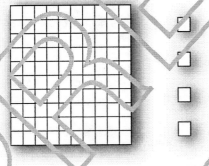

M2N1b

3. What is the difference?

$$70$$
$$-48$$

A 22

B 32

C 118

M2N2a

4. Meadow worked out this problem:

$56 + 39 = 95$

Which of these can she use to check her work?

A $95 - 39$

B $95 + 36$

C $56 - 39$

M2N2b

5. Which is the best estimate? Round to the nearest ten.

$$63$$
$$+12$$

A 50

B 70

C 75

M2N2e

6. What addition property is shown?

$78 + 84 = 84 + 78$

A Associative

B Commutative

C Identity

M2N2d

7. Which addition sentence relates to this multiplication sentence?

4×2

A $2 + 2 + 2 + 2 =$

B $8 + 4 + 2 =$

C $4 + 2 =$

M2N3a

8. Which array matches this multiplication sentence?

$6 \times 5 =$

A

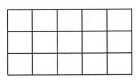

B

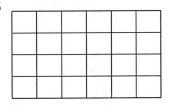

C

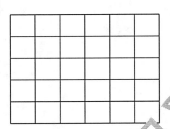

M2N3b

9. What is the product?

$9 \times 3 =$

A 8
B 18
C 27

M2N3b

10. How many equal groups of 3 can be made with the trophies below?

A 2
B 3
C 4

M2N3d

11. Which multiplication sentence matches the number line?

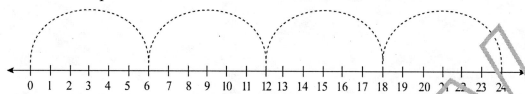

A $4 \times 4 =$
B $6 \times 6 =$
C $4 \times 6 =$

M2N3b

12. What is the missing factor?

$35 - 5 = 30$ $30 - 5 = 25$ $25 - 5 = 20$ $15 - 5 = 10$ $10 - 5 = 5$ $5 - 5 = 0$

_____ $\times 5 = 35$

A 3
B 5
C 7

M2N3

13. Which model shows an equal amount to the model below?

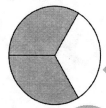

A

B

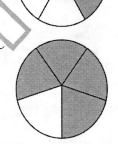

C

M2N4a

14. There are 6 juice bars in the freezer. All 6 juice bars are strawberry. Which fraction shows how many juice bars are strawberry?

A $\frac{3}{6}$

B $\frac{1}{6}$

C $\frac{6}{6}$

M2N4b

15. Temperature is measured in _____

A feet
B degrees
C minutes

M2M1c

16. Mr. Valez bakes 129 pizzas each Friday for students to purchase at the school cafeteria. He has baked 96 pizzas so far. How many more pizzas does he have left to bake?

A 33
B 96
C 129

M2N5b

17. What is the length to the nearest centimeter?

A 10 centimeters
B 11 centimeters
C 12 centimeters

M2M1a

18. Which length listed is the best estimate of the length of the shoe?

A   10 feet

B   10 inches

C   10 yards

19. Tristyn gives the clerk a one dollar bill to buy a pack of mints for $0.68. How much change should she receive?

A $0.32

B $0.68

C $1.68

20. What number should the minute hand be on when it is 9:00?

A   3

B   9

C   12

21. About what is the temperature?

   A 60 degrees

   B 75 degrees

   C 70 degrees

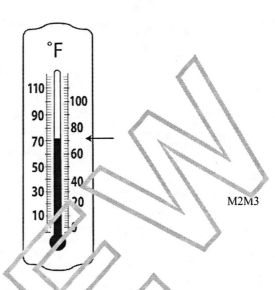

M2M3

22. How many vertices does this plane shape have?

   A 4
   B 5
   C 6

M2G1

23. How many edges does a trapezoid have?

   A 3
   B 4
   C 5

M2G1

24. What angle measures exactly 90°?

   A acute
   B obtuse
   C right

M2G1

25. How many edges does this solid have?

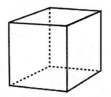

   A 6

   B 8

   C 12

26. What is the name of this solid?

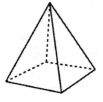

   A cone

   B cube

   C square pyramid

27. Of the solids below, which can be put together to form a cylinder?

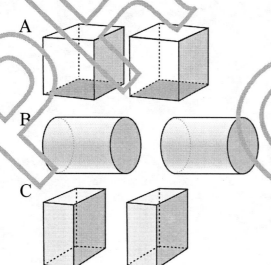

28. Which is the favorite place to swim?

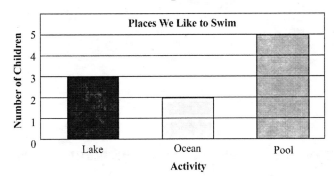

A lake
B ocean
C pool

M2D1b

29. How many children in total like grape and strawberry as favorite smoothies?

Smoothies We Like					
Banana	🍌	🍌	🍌		
Grape	🍇	🍇			
Strawberry	🍓	🍓	🍓	🍓	

Key: Each picture = 2 children

A 6
B 8
C 12

M2D1b

30. The Venn diagram shows different types of shapes. How many shapes are there in all?

A 12
B 5
C 4

M2D1b

Part 2

31. Which number sentence represents 929?

   A 92 hundreds + twenty-nine
   B 9 hundreds + nine
   C 9 hundreds + two tens + nine

M2N1a

32. Mr. Gomez's class read 790 books during the school year. Which figure models the number 790?

A

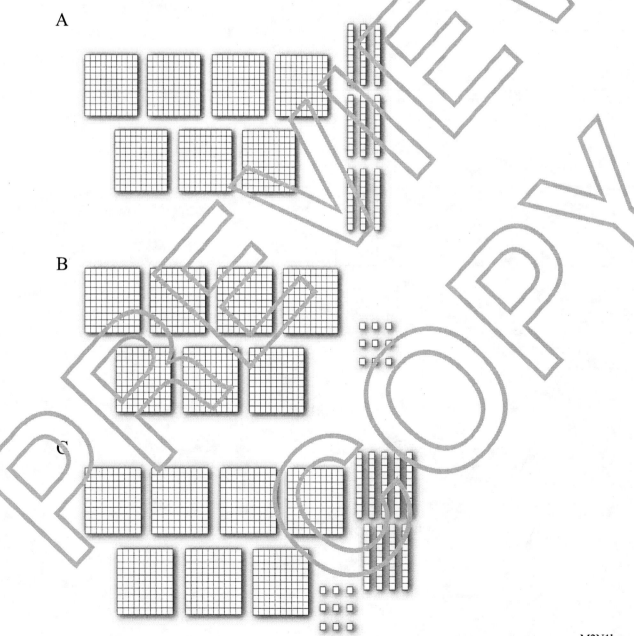

B

C

M2N1b

33. What is the sum?

$$725$$
$$+218$$

A 933

B 943

C 507

M2N2.

34. Dixon worked out this problem:

$88 + 44 = 134$

Which of these can he use to check his work?

A $88 - 44 =$

B $44 - 134 =$

C $134 - 44 =$

M2N 'b

35. Which is the best estimate? Round to the nearest ten.

$$72$$
$$-17$$

A 50

B 55

C 89

M2N2e

36. Which addition sentence relates to this multiplication sentence?

$6 \times 5 = 30$

A $5 + 6 =$

B $6 + 6 + 6 + 6 + 6 =$

C $5 + 5 + 5 + 5 + 5 =$

M2N3a

37. What is the product?

$8 \times 7 =$

A 15

B 54

C 56

M2N3b

38. Which multiplication sentence matches this array?

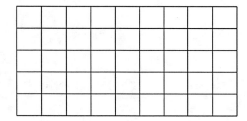

    A $5 \times 9 =$

    B $9 \times 9 =$

    C $5 \times 5 =$

39. How many equal groups of 6 can be made with award ribbons shown below?

    A 2

    B 3

    C 4

40. What is the missing factor?

$36 - 4 = 32$ $32 - 4 = 28$ $28 - 4 = 24$ $24 - 4 = 20$

$20 - 4 = 16$ $16 - 4 = 12$ $12 - 4 = 8$ $8 - 4 = 4$ $4 - 4 = 0$

    _____ $\times 4 = 36$

    A 7

    B 8

    C 9

41. Which fraction models the shaded part of the figure?

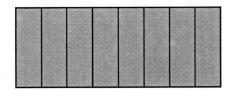

A $\dfrac{1}{8}$

B $\dfrac{4}{8}$

C $\dfrac{8}{8}$

M2N4b

42. Which shows $\dfrac{2}{3}$ of the fruit bar that is shaded?

A

B

C

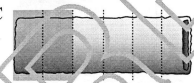

M2N4a

43. Which symbol should be used to compare the amounts?

A >

B <

C =

M2N5

44. Chris has 8 more grapes than Nicholas. Nicholas has 13 grapes. Which comparison matches the information?

  A $21 > 8 + 13$
  B $21 < 8 + 13$
  C $21 = 8 + 13$

M2N5

45. What is the missing value?

  $6 \times \boxed{} = 42$

  A 6
  B 7
  C 8

M2N5a

46. What is the length to the nearest inch?

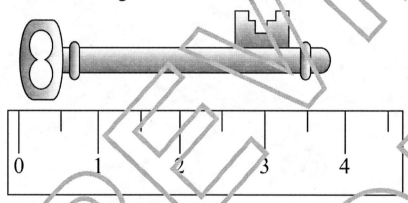

  A 2 inches
  B 3 inches
  C 4 inches

M2M1a

47. Which length listed is the best estimate of a banana?

  A 9 inches
  B 9 feet
  C 9 yards

M2M1b

48. Ashley bought beads for a necklace for $2.19 and a ribbon for $1.62. She gave the clerk $5.00. How much change should she receive back?

A $0.43

B $1.19

C $3.81

49. About what is the temperature?

A 25 degrees

B 35 degrees

C 45 degrees

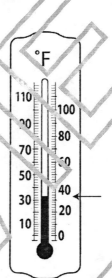

50. What time does the clock show?

A 6:21

B 7:00

C 7:21

51. What type of angle is shown?

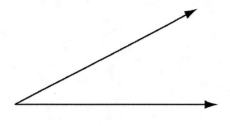

   A acute

   B right

   C obtuse

M2G1

52. Which plane shape listed has exactly 4 angles?

   A hexagon

   B pentagon

   C square

M2G1

53. How many edges does this plane shape have?

   A 4

   B 5

   C 6

M2G1

54. Which solid has both flat and curved surfaces?

   A cone

   B cube

   C pyramid

M2G2a

55. Which of these solids has 2 circle faces?

A

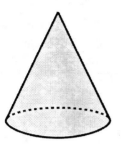

B

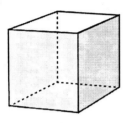

C

M2G2a

56. Which solid has only one vertex?

   A cone

   B cube

   C cylinder

M2G2b

57. What addition property is shown?

$(6 + 2) + 3 = 6 + (2 + 3)$

   A Associative

   B Commutative

   C Identity

M2N2d

58. How many children picked banana as a favorite smoothie?

Smoothies We Like					
Banana	🍌	🍌	🍌		
Grape	🍇	🍇			
Strawberry	🍓	🍓	🍓	🍓	

Key: Each picture = 2 children

A 4

B 5

C 6

M2D1b

59. How many children voted for an ocean as their favorite place to swim?

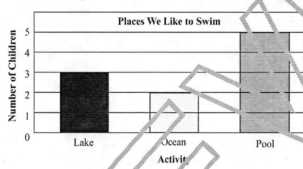

A 2

B 3

C 5

M2D1b

60. The Venn diagram shows different types of shapes. How many white triangles are there in all?

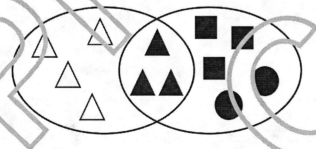

A 3

B 4

C 7

M2D1b

Practice Test 2

Part 1

1. Leo counted 321 pages in the book. Which figure models the number 321?

A

B

C

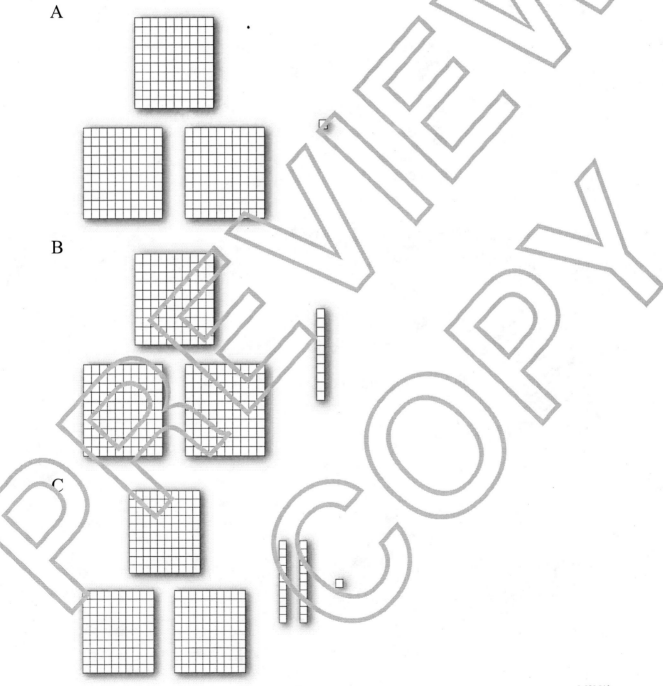

2. Which number matches 8 hundreds + 2 tens + nine?

  A 829
  B 8,209
  C 8,029

M2N1a

3. What is the difference?

$$\begin{array}{r} 60 \\ -41 \\ \hline \end{array}$$

  A 19
  B 20
  C 101

M2N2a

4. Luke worked out this problem:

$79 + 15 = 94$

Which of these can he use to check his work?

  A $94 + 79$
  B $79 - 15$
  C $94 - 15$

M2N2b

5. Which is the best estimate? Round to the nearest ten.

$$\begin{array}{r} 56 \\ +19 \\ \hline \end{array}$$

  A 40
  B 70
  C 80

M2N2e

6. What is the addition property shown?

$213 + 98 = 98 + 213$

  A Associative
  B Commutative
  C Identity

M2N2d

7. Which addition sentence relates to this multiplication sentence?

$7 \times 4 = 28$

A $7 + 7 + 7 + 7 =$

B $7 + 4 + 28 =$

C $4 + 7 =$

8. Which tool would you use to measure the temperature of the water in a pool?

A thermostat

B barometer

C thermometer

9. What is the product?

$8 \times 4 =$

A 16

B 24

C 32

10. How many equal groups of 4 can be made with the flowers below?

A 3

B 5

C 6

11. Which multiplication sentence matches the number line?

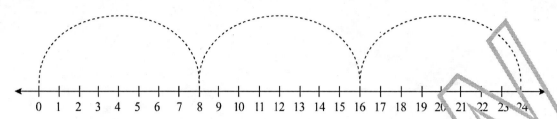

A $3 \times 3 =$
B $8 \times 3 =$
C $8 + 8 + 8 =$

M2N3b

12. What is the missing factor?

$8 - 2 = 6$ $6 - 2 = 4$ $4 - 2 = 2$ $2 - 2 = 0$

_____ $\times 2 = 8$

A 3
B 4
C 5

M2N3d

13. There are 5 birds in the tree. All 5 birds are brown. Which fraction shows how many birds are brown?

A $\dfrac{1}{10}$

B $\dfrac{5}{5}$

C $\dfrac{1}{5}$

M2N4b

14. Which symbol should be used to compare the numbers?

642 ☐ 642

A >
B <
C =

M2N5

15. Which model shows an equal amount to the model below?

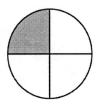

A
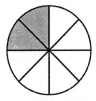

B

C
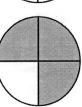

M2N4a

16. Mr. Jackson runs 100 laps each week. He has run 56 laps so far. How many more laps does he have left?

A 44

B 54

C 156

M2N5b

17. What is the length to the nearest centimeter?

A 8 centimeters

B 9 centimeters

C 10 centimeters

M2M1a

18. Which length listed is the best estimate of a mat?

A 34 inches

B 34 feet

C 34 yards

M2M1b

19. Nicholas gives the clerk two dollar bills to buy a drink for $1.18. How much change should he receive?

A $0.18

B $0.22

C $0.82

M2N1c

20. What number should the minute hand be on when it is 7:55?

A 5

B 7

C 11

M2M2

21. About what is the temperature?

A 25 degrees

B 35 degrees

C 30 degrees

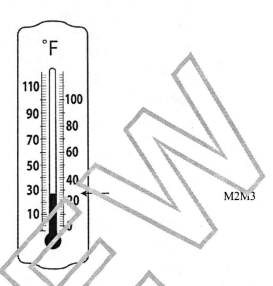

M2M3

22. How many vertices does this plane shape have?

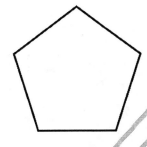

A 3

B 4

C 5

M2G1

23. How many edges does a right triangle have?

A 3

B 4

C 5

M2G1

24. A right angle measures exactly _____

A 30°

B 60°

C 90°

M2G1

25. How many faces does this solid have?

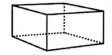

A 12
B 8
C 6

26. Which solid can roll and slide?

A cylinder
B pyramid
C rectangular prism

27. Which new solid can be made with these two solids?

A cone
B sphere
C cylinder

28. What is the difference between the most favorite place to swim and the least favorite place to swim?

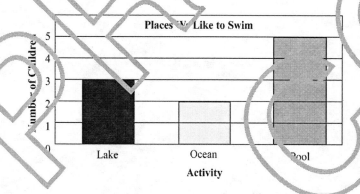

A 5
B 3
C 2

190

29. What is the difference between children that like banana smoothies and children that like grape smoothies?

Smoothies We Like					
Banana	🍌	🍌	🍌		
Grape	🍇	🍇			
Strawberry	🍓	🍓	🍓	🍓	

Key: Each picture = 2 children

A 1
B 2
C 5

M2D1b

30. The Venn diagram shows different types of shapes. How many white shapes are there in all?

A 3
B 4
C 5

M2D1b

Part 2

31. Mrs. Whitakar's class read 810 books during the school year. Which figure models the number 810?

A

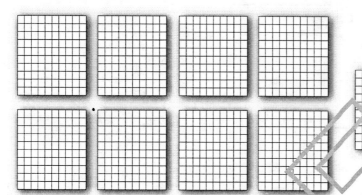

B

C

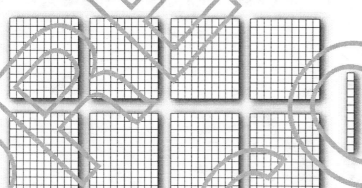

32. Which number sentence represents 412?

   A $412 + 12$

   B $400 + 10 + 2$

   C $400 + 12 + 2$

33. What is the sum?

$$807 \\ +194$$

A $1,101$

B $1,001$

C 992

34. Asia worked out this problem:

$26 + 72 = 98$

Which of these can she use to check her work?

A $98 - 26$

B $72 - 26$

C $72 - 98$

35. What addition property is shown?

$(26 + 21) + 37 = 26 + (21 + 37)$

A Associative

B Commutative

C Identity

36. Which is the best estimate? Round to the nearest ten.

$$81 \\ -12$$

A 93

B 70

C 69

37. Which addition sentence relates to this multiplication sentence?

$6 \times 7 = 42$

A $6 + 7 =$

B $7 + 7 + 7 + 7 + 7 =$

C $6 + 6 + 6 + 6 + 6 + 6 + 6 =$

38. What is the product?

$9 \times 6 =$

A 15
B 54
C 56

M2N3b

39. Which multiplication sentence matches this array?

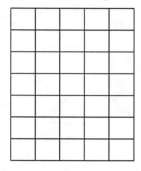

A $7 \times 5 =$
B $7 \times 7 =$
C $5 \times 5 =$

M2N3b

40. What is the missing factor?

$32 - 4 = 28$　　$28 - 4 = 24$　　$24 - 4 = 20$　　$21 - 4 = 16$
$16 - 4 = 12$　　$12 - 4 = 8$　　$8 - 4 = 4$　　$4 - 4 = 0$

_____ $\times 4 = 32$

A 6
B 7
C 8

M2N3d

41. Which multiplication sentence matches the number line?

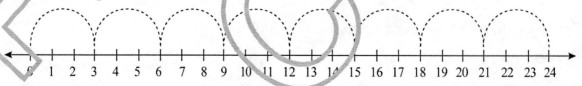

A $3 \times 3 =$
B $24 \times 3 =$
C $3 \times 8 =$

M2N3b

42. Which fraction represents the shaded part of the model?

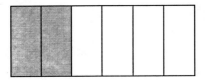

A $\dfrac{2}{6}$

B $\dfrac{4}{6}$

C $\dfrac{6}{6}$

M2N4b

43. Chandler ate $\dfrac{1}{3}$ of the granola bar. Which shows $\dfrac{1}{3}$ of the granola bar that is shaded?

A

B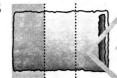

C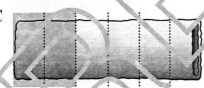

M2N4a

44. Which symbol should be used to compare the amounts?

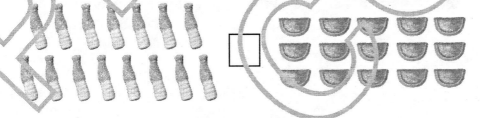

A >

B <

C =

M2N5

45. Joe has 4 more crayons than Steve. Steve has 15 crayons. Which comparison matches the information?

A $15 > 4 + 15$
B $15 = 4 + 15$
C $15 < 4 + 15$

46. What is the missing value?

$9 \times \boxed{} = 72$

A 6
B 7
C 8

47. What is the length to the nearest inch?

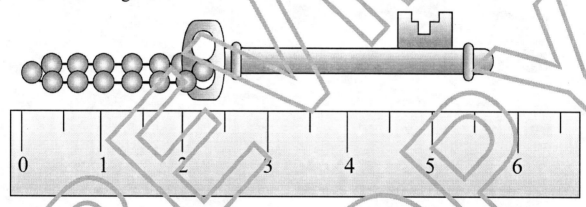

A 5 inches
B 6 inches
C 7 inches

48. Which length listed as the best estimate of the length of a remote control?

A 7 inches
B 7 feet
C 7 yards

49. Kaitlyn bought a pen for $2.05 and a pad for $1.99. She gave the clerk $5.00. How much change should she receive back?

A $4.04
B $2.95
C $0.96

50. About what is the temperature?

A 75 degrees

B 85 degrees

C 80 degrees

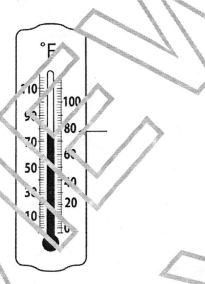

51. What time does the clock show?

A 5:05
B 5:25
C 5:28

52. Which plane shape listed has exactly 5 angles?

A hexagon
B pentagon
C square

53. What type of angle is shown?

    A acute
    B right
    C obtuse

54. Which of these solids has 6 square faces?

    A

    B

    C

55. How many circle faces are on a cylinder?

    A 1
    B 2
    C 3

56. Which solid has only flat surfaces?

    A cone
    B cylinder
    C pyramid

57. Which solid has five vertices?

    A cone
    B cube
    C square pyramid

58. How many children did NOT pick strawberry as a favorite smoothie?

Smoothies We Like				
Banana	🍌	🍌	🍌	
Grape	🍇	🍇		
Strawberry	🍓	🍓	🍓	🍓

Key: Each picture = 2 children

A 5

B 8

C 10

59. How many children voted for an ocean and a pool as a favorite place to swim?

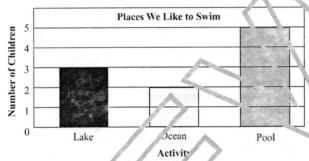

A 3

B 5

C 7

60. The Venn diagram shows different types of shapes. How many triangles are NOT white?

A 3

B 4

C 5

Index

CRCT

Please fill out the form completely, and return by mail or fax to American Book Company.

Purchase Order #: _____   Date: _____

Contact Person: _____

School Name (and District, if any): _____

Billing Address: _____   Street Address: _____ ☐ same as billing

Attn: _____   Attn: _____

Phone: _____   E-Mail: _____

Credit Card #: _____   Exp Date: _____

Authorized Signature: _____

Order Number	Product Title	Pricing* (10 books)	Qty	Pricing (30+ books)	Qty	Total Cost
GA1-R0409	Mastering the Georgia 1st Grade CRCT in Reading	$169.90 (1 set of 10 books)		$329.70 (1 set of 30 books)		
GA2-M0409	Mastering the Georgia 2nd Grade CRCT in Science	$169.90 (1 set of 10 books)		$329.70 (1 set of 30 books)		
GA2-H0409	Our State of Georgia (2nd Grade Social Studies)	$169.90 (1 set of 10 books)		$329.70 (1 set of 30 books)		
GA3-M0607	Mastering the Georgia 3rd Grade CRCT in Math	$169.90 (1 set of 10 books)		$329.70 (1 set of 30 books)		
GA3-R0607	Mastering the Georgia 3rd Grade CRCT in Reading	$169.90 (1 set of 10 books)		$329.70 (1 set of 30 books)		
GA3-S0508	Mastering the Georgia 3rd Grade CRCT in Science	$169.90 (1 set of 10 books)		$329.70 (1 set of 30 books)		
GA3-H1008	Mastering the Georgia 3rd Grade CRCT in Social Studies	$169.90 (1 set of 10 books)		$329.70 (1 set of 30 books)		
GA4-M0808	Mastering the Georgia 4th Grade CRCT in Math	$169.90 (1 set of 10 books)		$329.70 (1 set of 30 books)		
GA4-R0808	Mastering the Georgia 4th Grade CRCT in Reading	$169.90 (1 set of 10 books)		$329.70 (1 set of 30 books)		
GA4-S0708	Mastering the Georgia 4th Grade CRCT in Science	$169.90 (1 set of 10 books)		$329.70 (1 set of 30 books)		
GA4-H1008	Mastering the Georgia 4th Grade CRCT in Social Studies	$169.90 (1 set of 10 books)		$329.70 (1 set of 30 books)		
GA5-M0806	Mastering the Georgia 5th Grade CRCT in Math	$169.90 (1 set of 10 books)		$329.70 (1 set of 30 books)		
GA5-R1206	Mastering the Georgia 5th Grade CRCT in Reading	$169.90 (1 set of 10 books)		$329.70 (1 set of 30 books)		
GA5-S1107	Mastering the Georgia 5th Grade CRCT in Science	$169.90 (1 set of 10 books)		$329.70 (1 set of 30 books)		
GA5-H0808	Mastering the Georgia 5th Grade CRCT in Social Studies	$169.90 (1 set of 10 books)		$329.70 (1 set of 30 books)		
GA5-W1008	Mastering the Georgia Grade 5 Writing Assessment	$169.90 (1 set of 10 books)		$329.70 (1 set of 30 books)		
GA6-L0508	Mastering the Georgia 6th Grade CRCT in ELA	$169.90 (1 set of 10 books)		$329.70 (1 set of 30 books)		
GA6-M0305	Mastering the Georgia 6th Grade CRCT in Math	$169.90 (1 set of 10 books)		$329.70 (1 set of 30 books)		
GA6-R0108	Mastering the Georgia 6th Grade CRCT in Reading	$169.90 (1 set of 10 books)		$329.70 (1 set of 30 books)		
GA6-S1206	Mastering the Georgia 6th Grade CRCT in Science	$169.90 (1 set of 10 books)		$329.70 (1 set of 30 books)		
GA6-H0208	Mastering the Georgia 6th Grade CRCT in Social Studies	$169.90 (1 set of 10 books)		$329.70 (1 set of 30 books)		
GA7-L0508	Mastering the Georgia 7th Grade CRCT in ELA	$169.90 (1 set of 10 books)		$329.70 (1 set of 30 books)		
GA7-M0305	Mastering the Georgia 7th Grade CRCT in Math	$169.90 (1 set of 10 books)		$329.70 (1 set of 30 books)		
GA7-R0707	Mastering the Georgia 7th Grade CRCT in Reading	$169.90 (1 set of 10 books)		$329.70 (1 set of 30 books)		
GA7-S1206	Mastering the Georgia 7th Grade CRCT in Science	$169.90 (1 set of 10 books)		$329.70 (1 set of 30 books)		
GA7-H0208	Mastering the Georgia 7th Grade CRCT in Social Studies	$169.90 (1 set of 10 books)		$329.70 (1 set of 30 books)		
GA8-L0505	Passing the Georgia 8th Grade CRCT in ELA	$169.90 (1 set of 10 books)		$329.70 (1 set of 30 books)		
GA8-MATH08	Passing the Georgia 8th Grade CRCT in Math	$169.90 (1 set of 10 books)		$329.70 (1 set of 30 books)		
GA8-R0505	Passing the Georgia 8th Grade CRCT in Reading	$169.90 (1 set of 10 books)		$329.70 (1 set of 30 books)		
GA8-S0707	Passing the Georgia 8th Grade CRCT in Science	$169.90 (1 set of 10 books)		$329.70 (1 set of 30 books)		
GA8-H0607	Passing the Georgia 8th Grade CRCT in Georgia Studies	$169.90 (1 set of 10 books)		$329.70 (1 set of 30 books)		
GA8-W0907	Passing the Georgia Grade 8 Writing Assessment	$169.90 (1 set of 10 books)		$329.70 (1 set of 30 books)		

1-5-09   *Minimum order is 1 set of 10 books of the same subject.

Subtotal

Shipping & Handling 12%

Total

American Book Company ● PO Box 2638 ● Woodstock, GA 30188-1383
Toll Free Phone: 1-888-264-5877 ● Toll-Free Fax: 1-866-827-3240
Web Site: www.americanbookcompany.com

Call Toll-Free 1-888-264-5877 to ORDER and for FREE PREVIEW COPIES!

Visit americanbookcompany.com to download FREE SAMPLES of all of our products!